# Introduction

How often do we hear people saying things like "I hate math", "Numbers are not my thing", and "Can I use a calculator for that"? Nowadays, many people disregard the value of numbers, thinking they can live their lives free of the numbers that once haunted them in grade school. However, dealing with numbers should not be intimidating. Mathematics and numbers, dictate a huge portion of our everyday lives, and it is time that we began to see numbers in a new light – as a friend, not a foe!

Numerical puzzles are excellent educational fun. In this brand new puzzle book, we attempt to combine numbers, graph theory, and geometry. This work is based on the very popular five pronged star graph. The book contains 400 numerically different puzzles of the same type. Each page has 4 puzzles, and there are 100 pages of puzzles. There are five numbers on the sides of every star, and the player needs to figure out five numbers to put on the empty vertices of the star to solve the puzzle.

There is a unique solution to every puzzle. I.e. There are 5 numbers that if put correctly at the vertices of the star, the puzzle will be solved correctly. This situation is similar to our previous work "The High 5". The twist in this new book is the direction of the straight segments of the graph which adds a new challenge and takes more focusing to solve. Only positive integers/ whole numbers are used in this book and in solutions. First, the player has to start by choosing which side of the star to decompose into two numbers. Then, using the intial guess, the player can solve for the rest of the numbers on the vertices. If a players initial guess was incorrect, the player has to start over again. Trial and error is a good beginner strategy, so players are recommended to use pencil and eraser to easily fix their mistakes. There are no answers given at the end of the book because every exercise is easily self-proved. By the time a player solves a few of these puzzles, he/she will develop a sense of strategy and method to reach a solution faster.

Teachers and students can use this book for contests, or to inspire learning. It can also be used as an activity book, or a work book. This book is great for math and puzzle lovers. Try it timed for an extra challenge! Share it with family and friends, and spread the word!

Next page for instructions

# INSTRUCTIONS

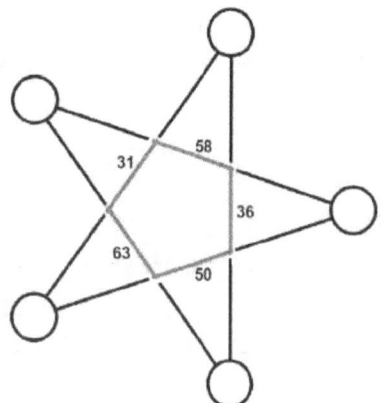

Pick a number on one of the straight lines of the inner pentagon (highlighted in red). Separate the number into two numbers that add to the number on the line.

For example, 36 can be summed by 18 + 18.

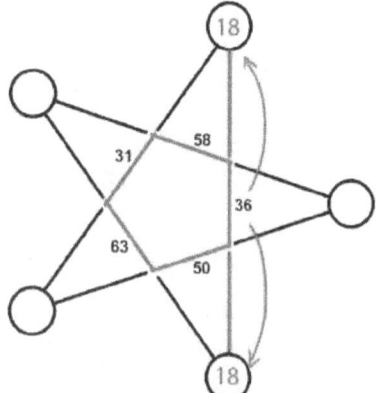

Place the two numbers you have chosen on the circles that lie on the endpoints of the line.

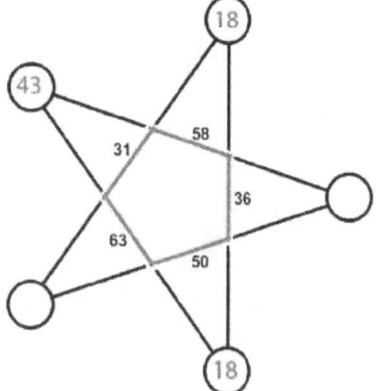

Using the information you have, you can easily solve for the rest of the numbers in the circles.

Given the 18 in the bottommost circle, you can solve for the top left circle.
18 + x = 63
x = 61 - 18
x = 43

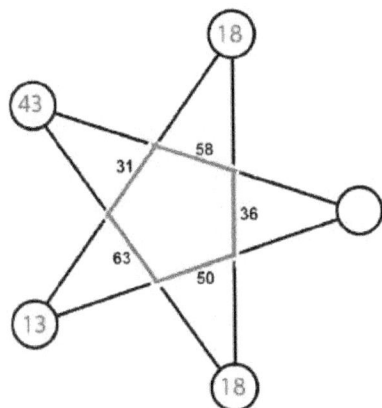

You can now solve for the two numbers that add to 31.
18 + x = 31
x = 31 - 18
x = 13

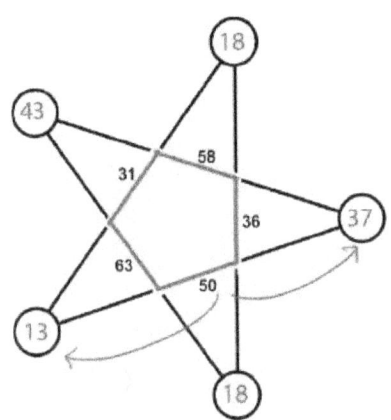

Finally, you can solve for the last circle. If your initial two numbers were corect, the puzzle should be completely solved.

The last circle creates the following equation:
13 + x = 50
x = 50 - 13
x = 37

HOWEVER, because 37 + 43 does NOT equal 58, the initial two numbers of 18 and 18 were INCORRECT. The puzzle has not been solved yet! You have to try two new numbers. If you find the correct two numbers, the entire puzzle should fit together perfectly, and each number will add exactly to the numbers on the lines of the pentagon.

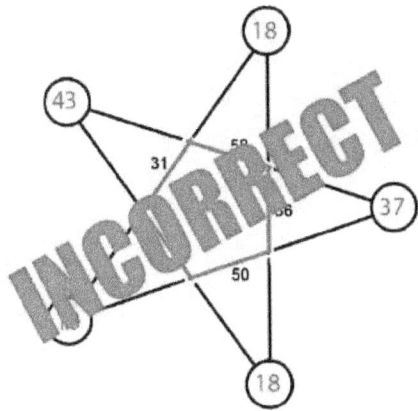

Lets try one more time!

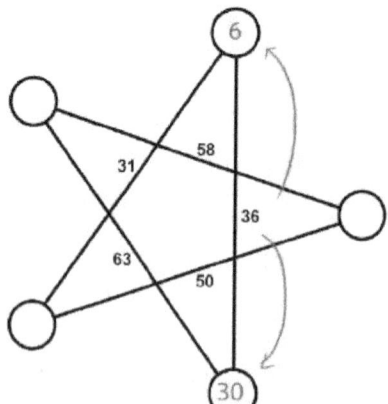

Lets try the numbers 30 and 6.
30 + 6 = 36

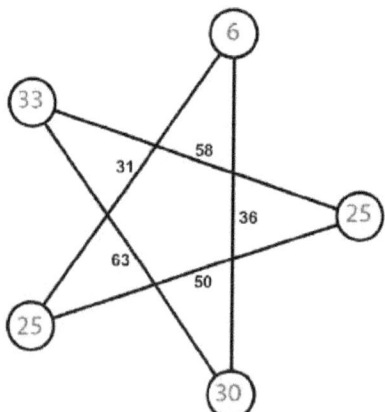

Using these two numbers, you can solve for the rest of the numbers on the vertices star.

x + 30 = 63
x = 33

y + 33 = 58
y = 25

z + 25 = 50
z = 25

This is the correct solution!
You can prove this easily:
58 = 25 + 33
36 = 30 + 6
50 = 25 + 25
63 = 33 + 30
31 = 6 + 25

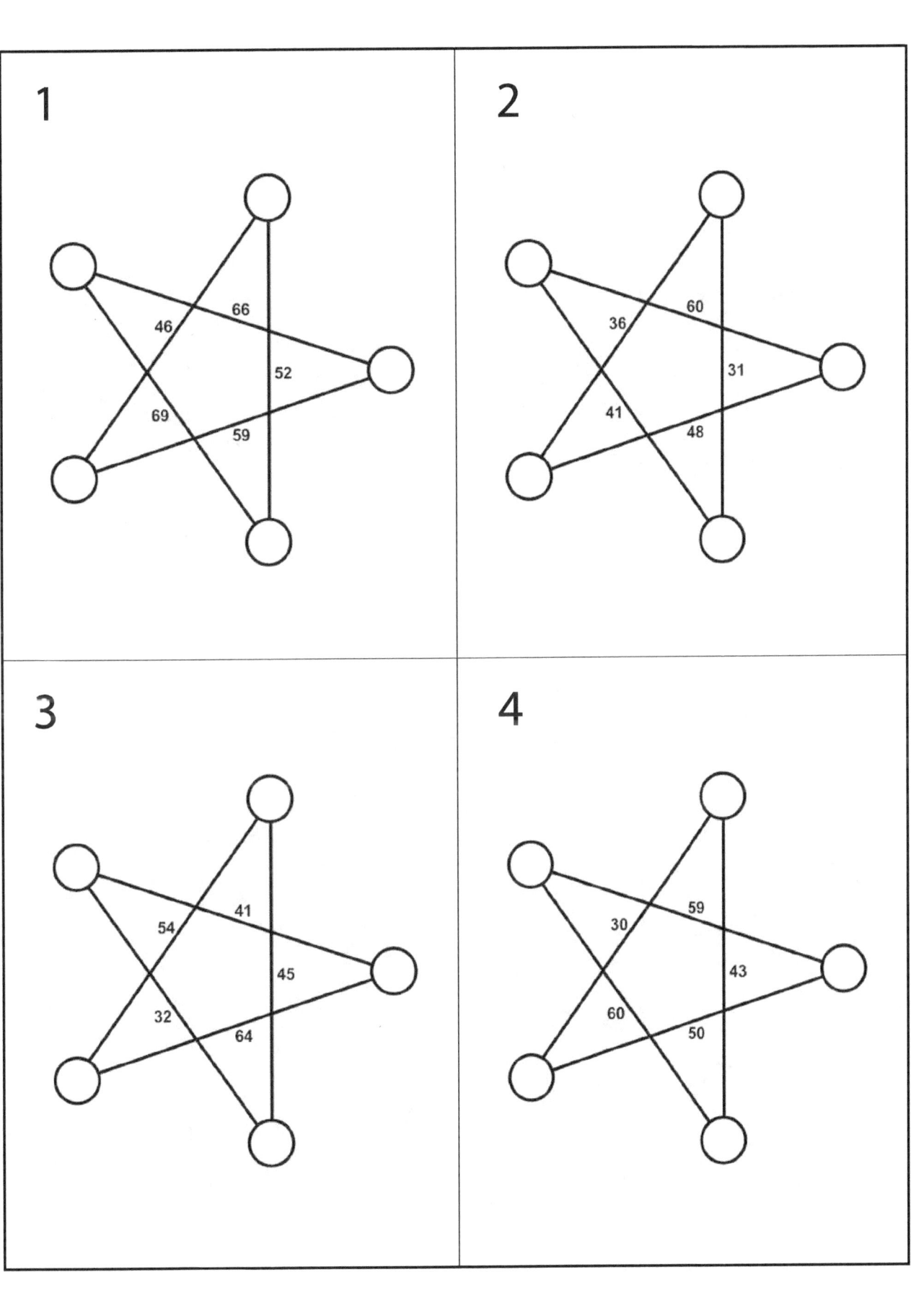

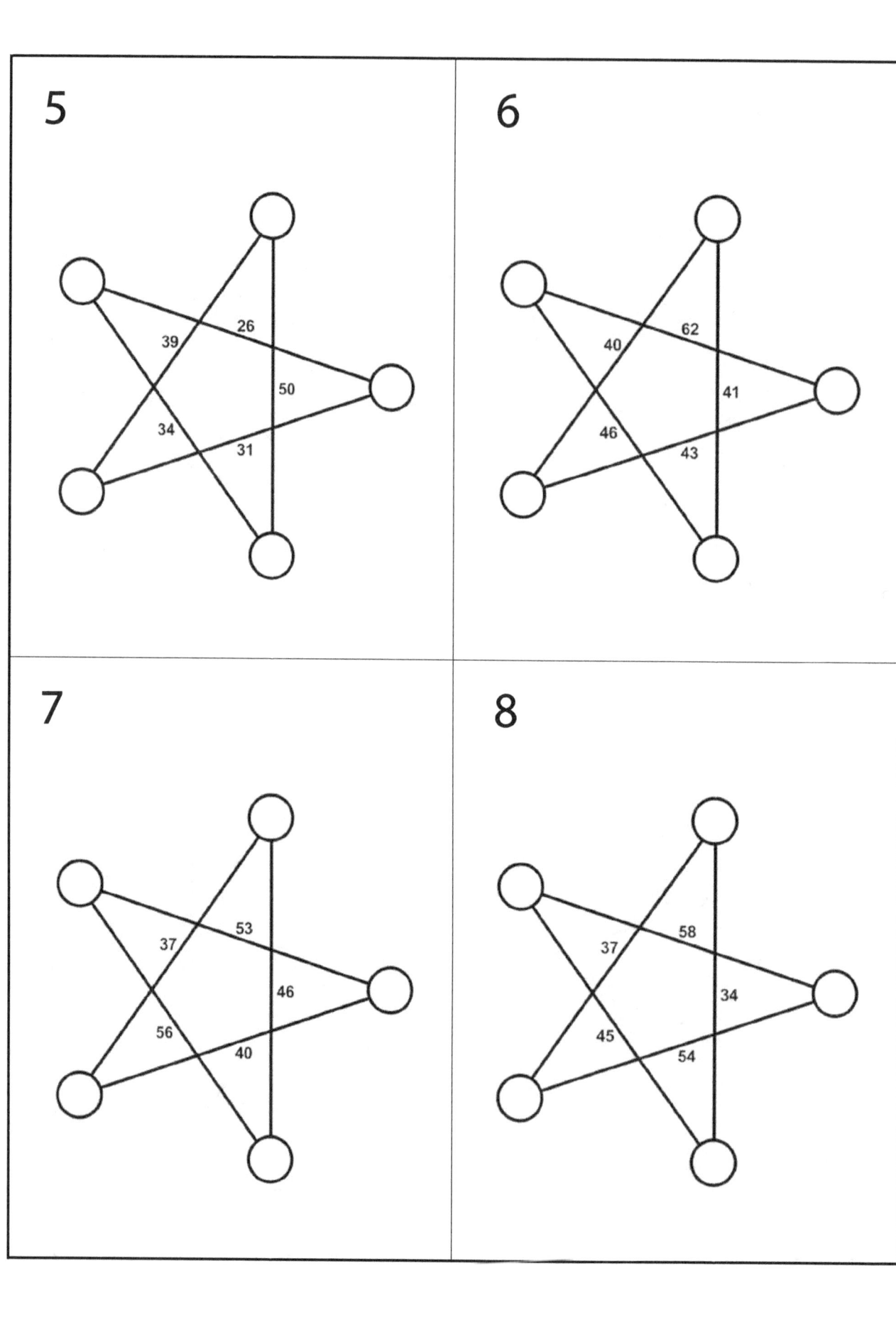

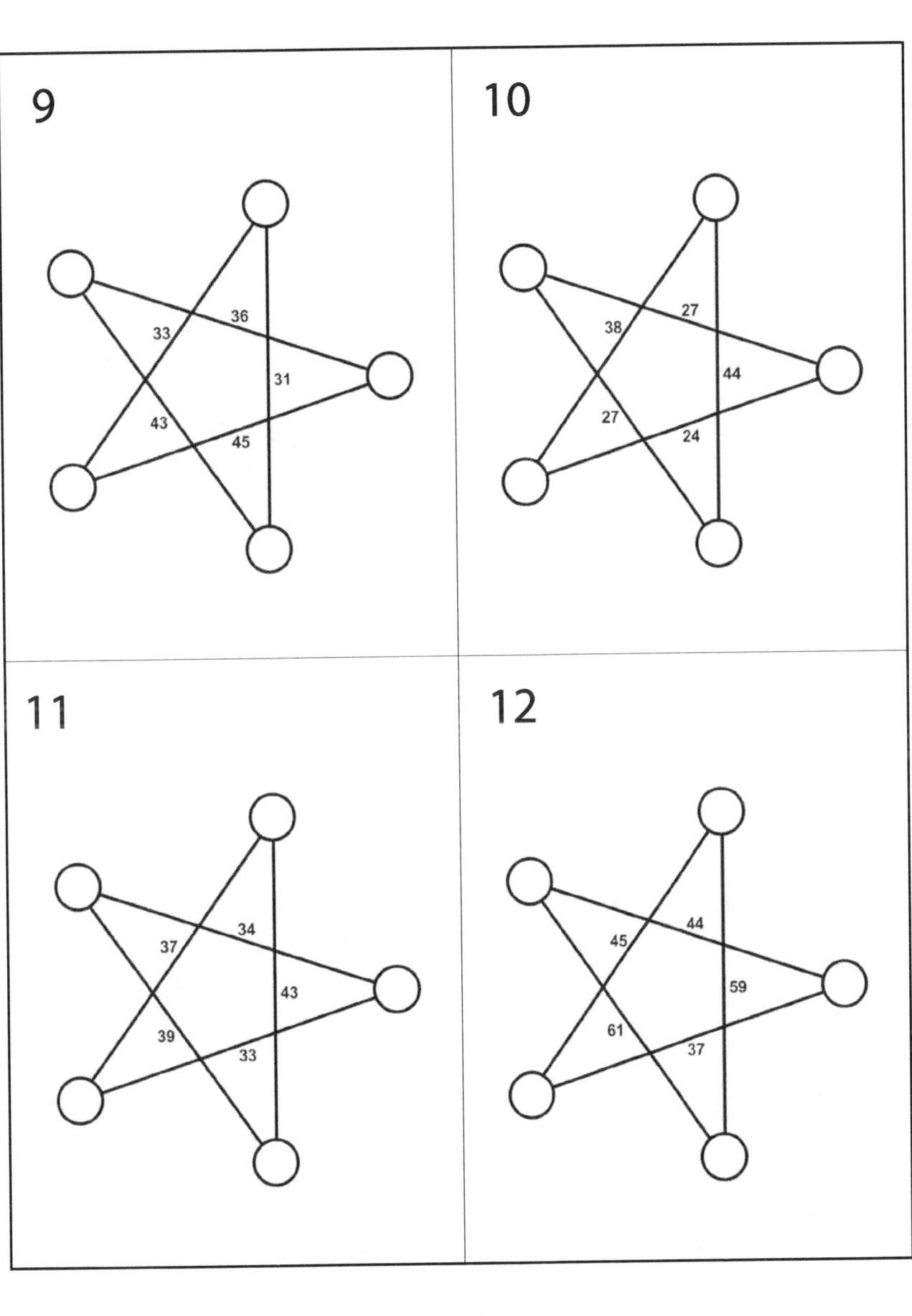

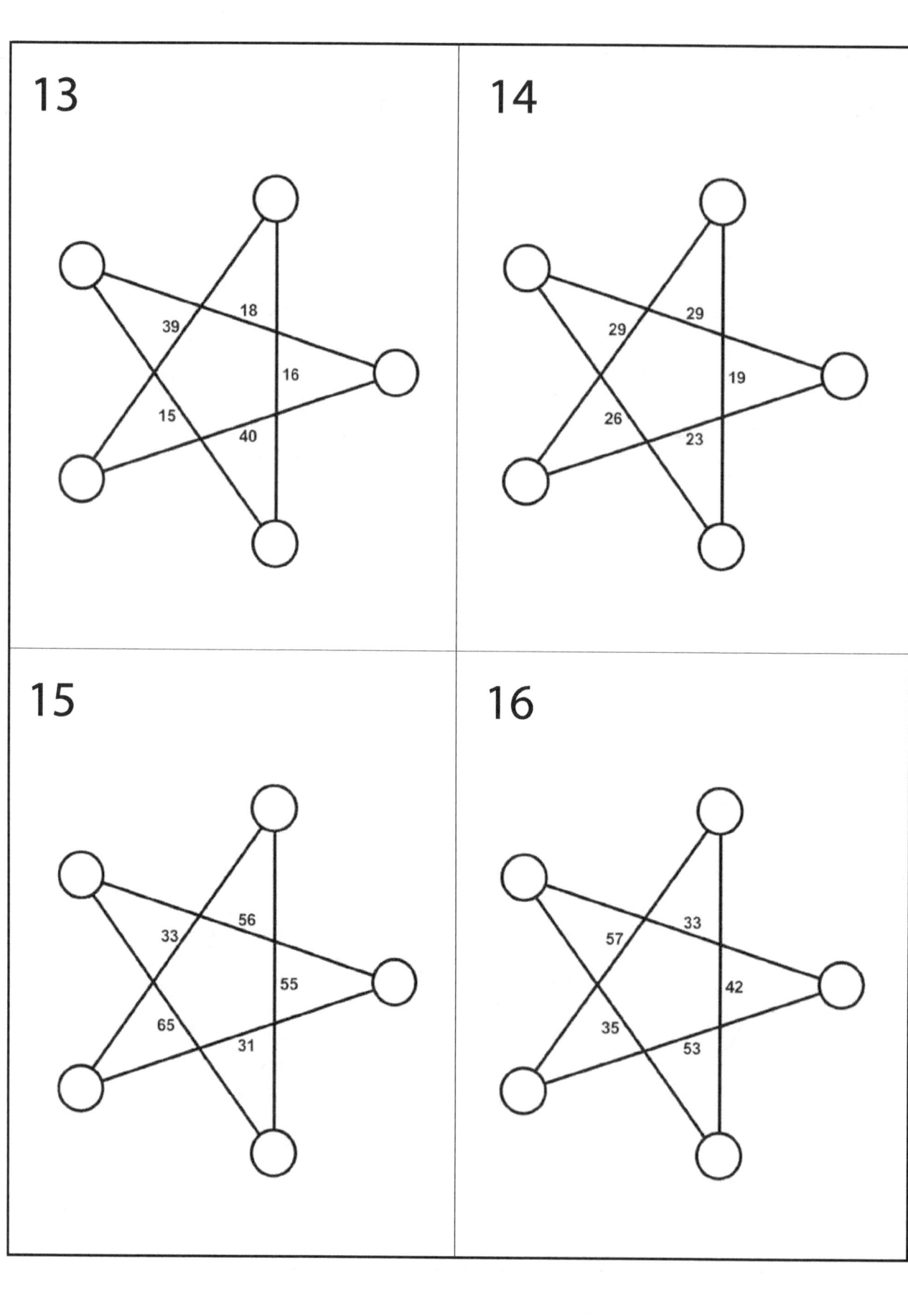

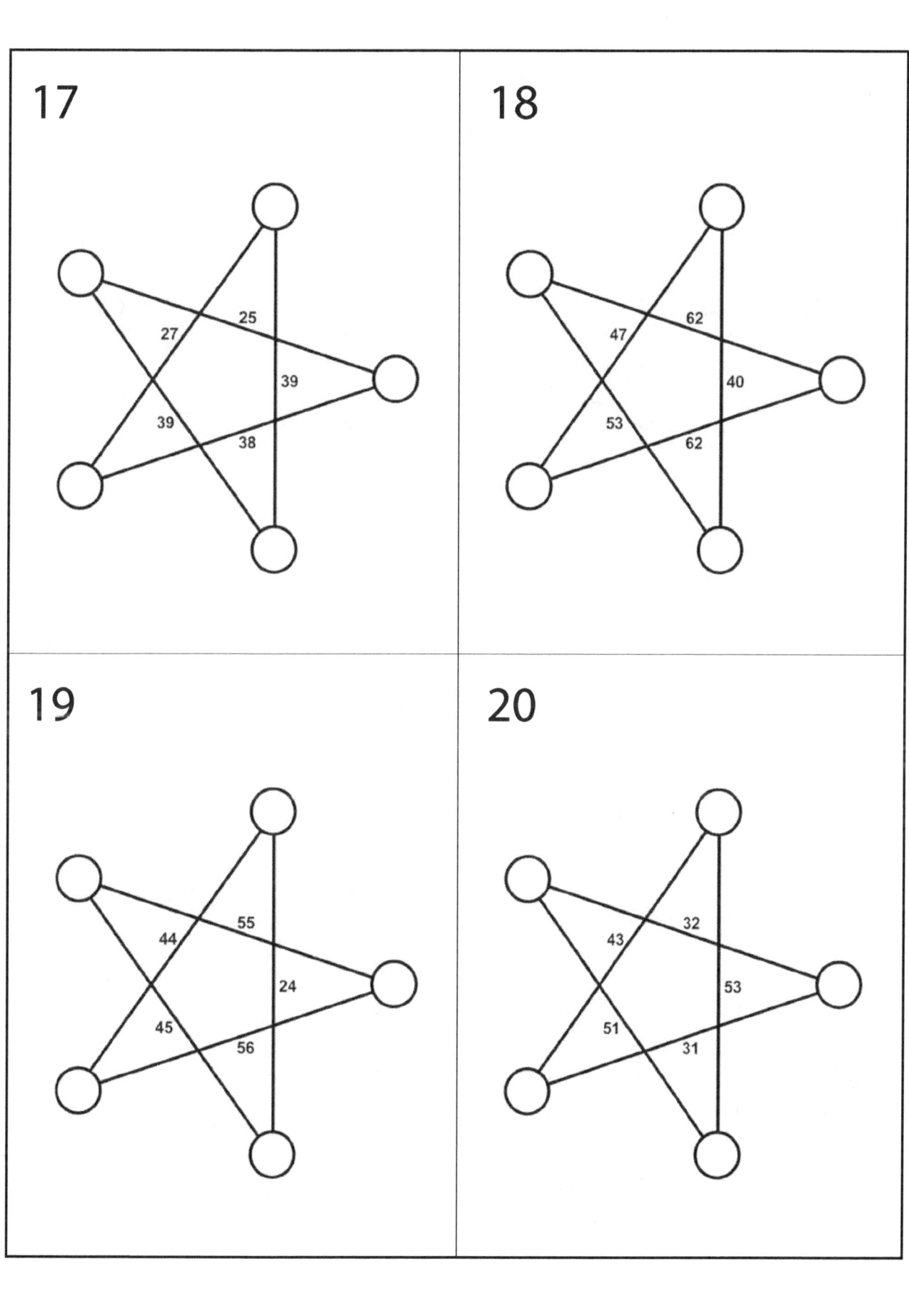

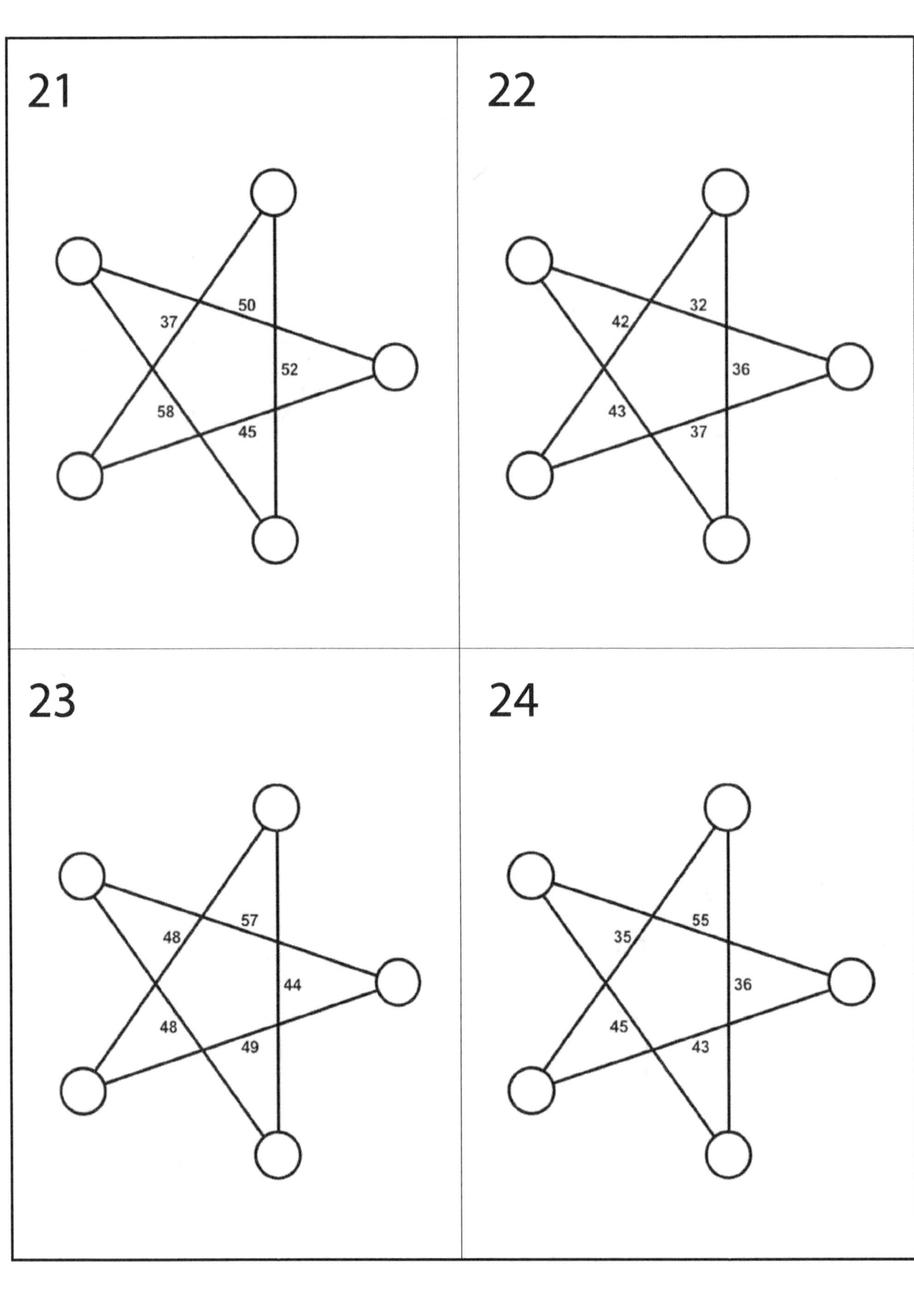

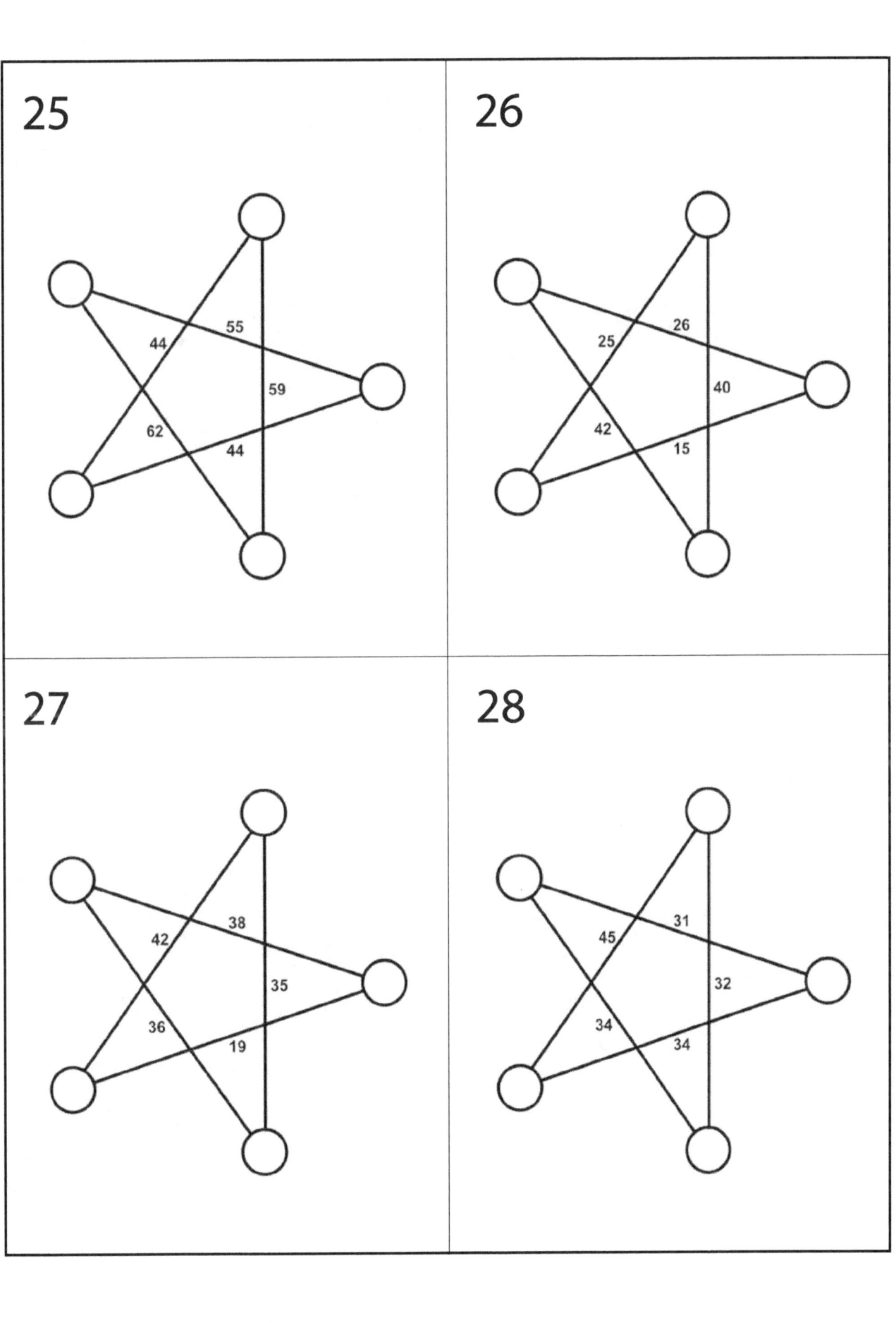

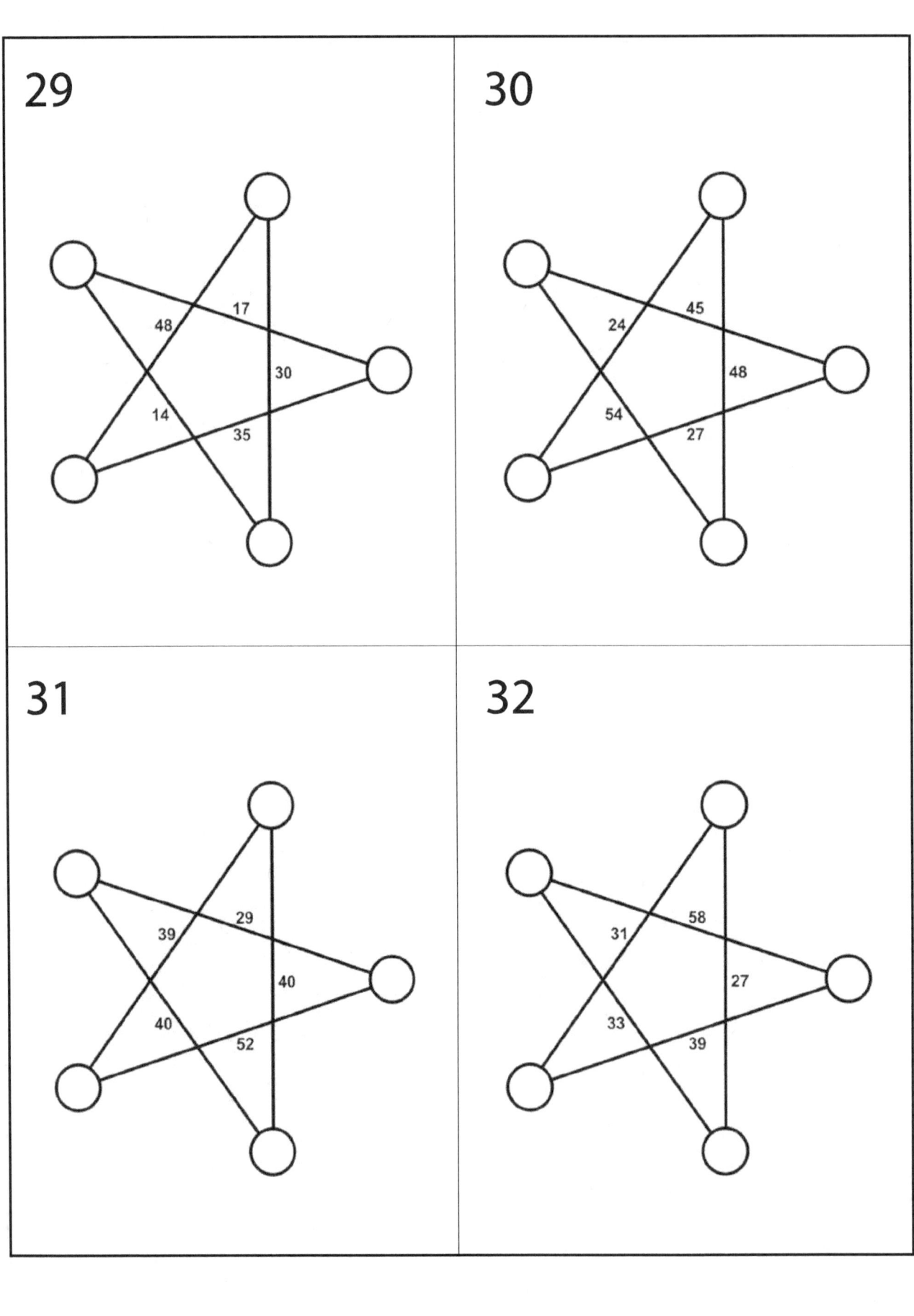

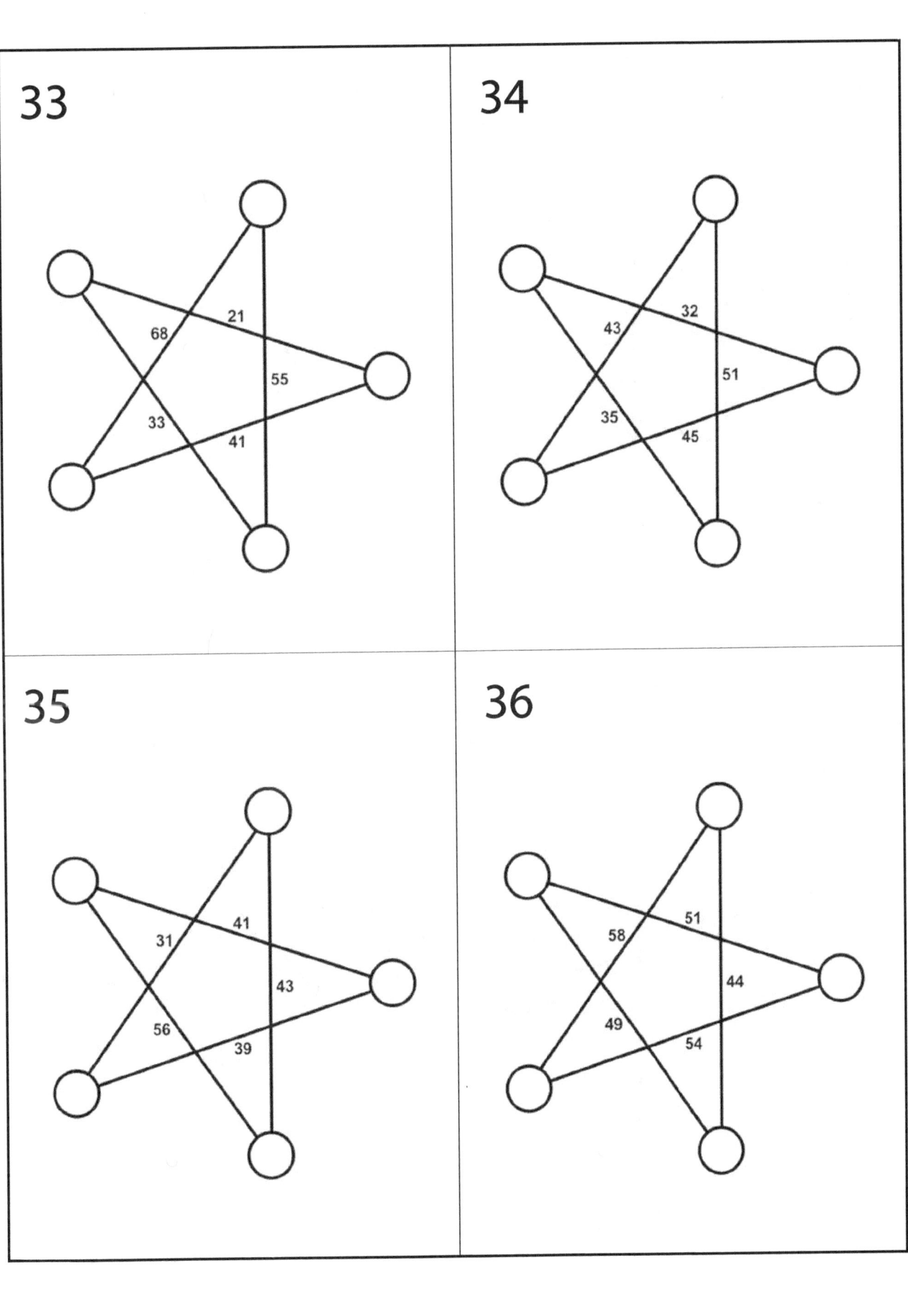

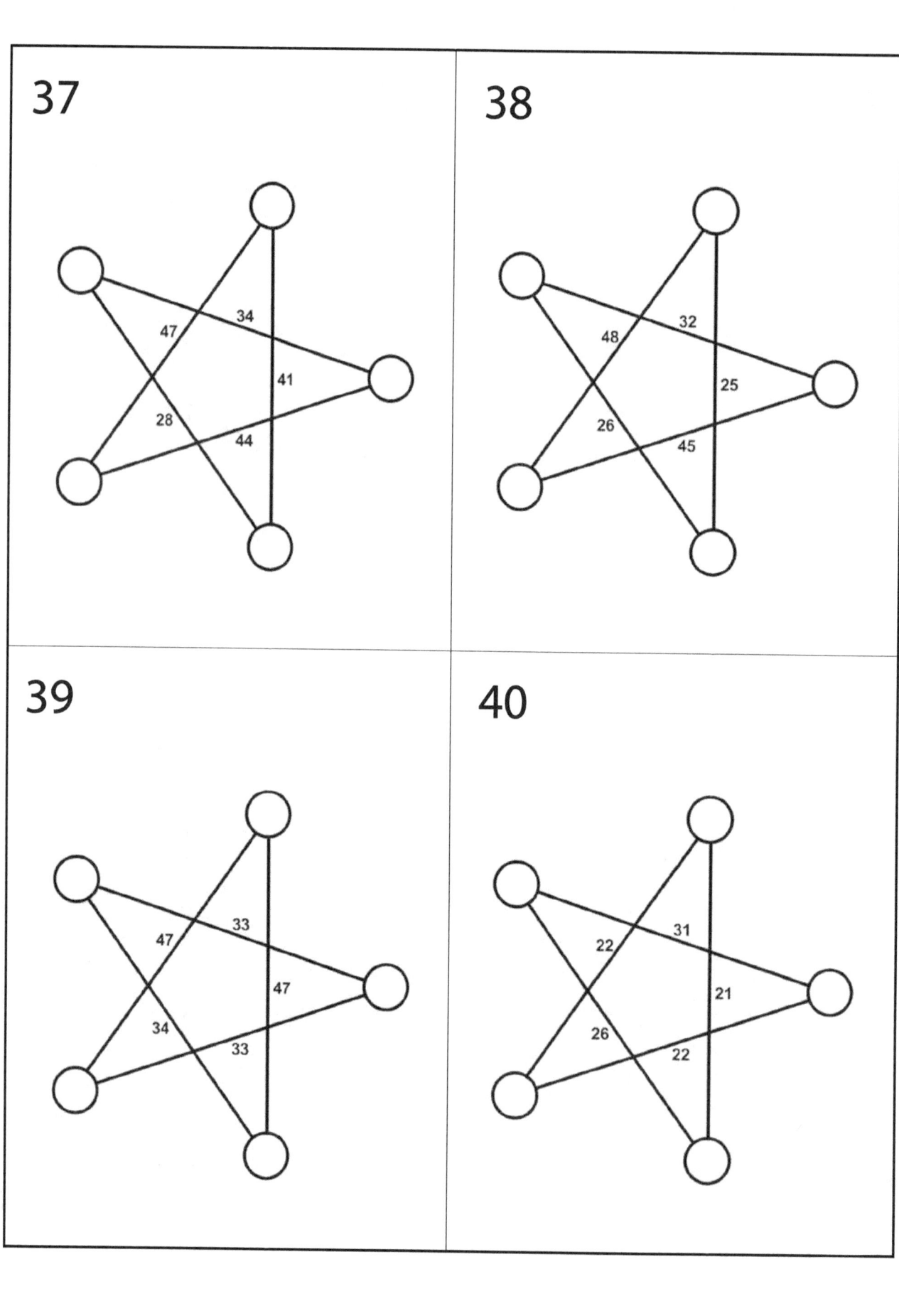

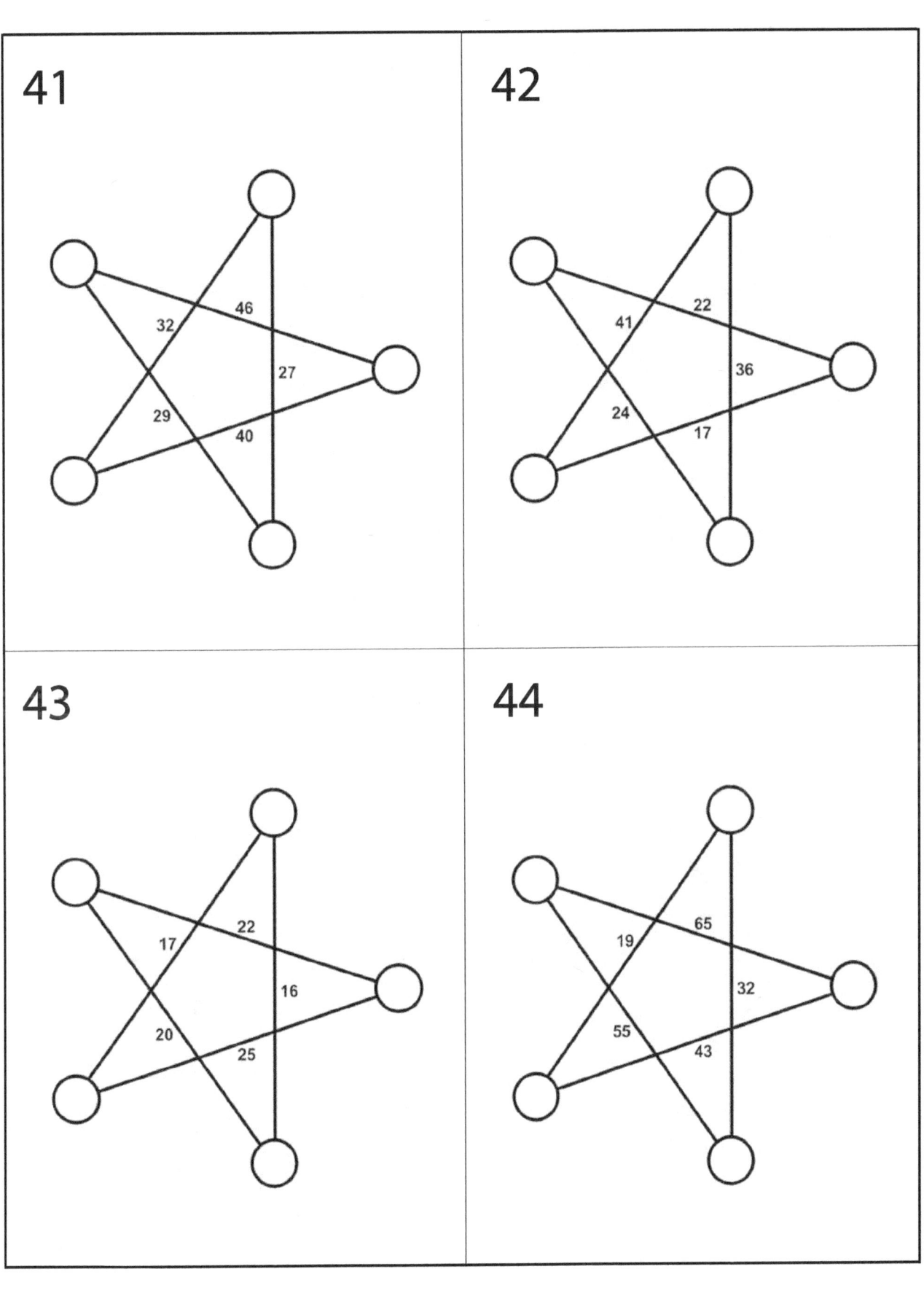

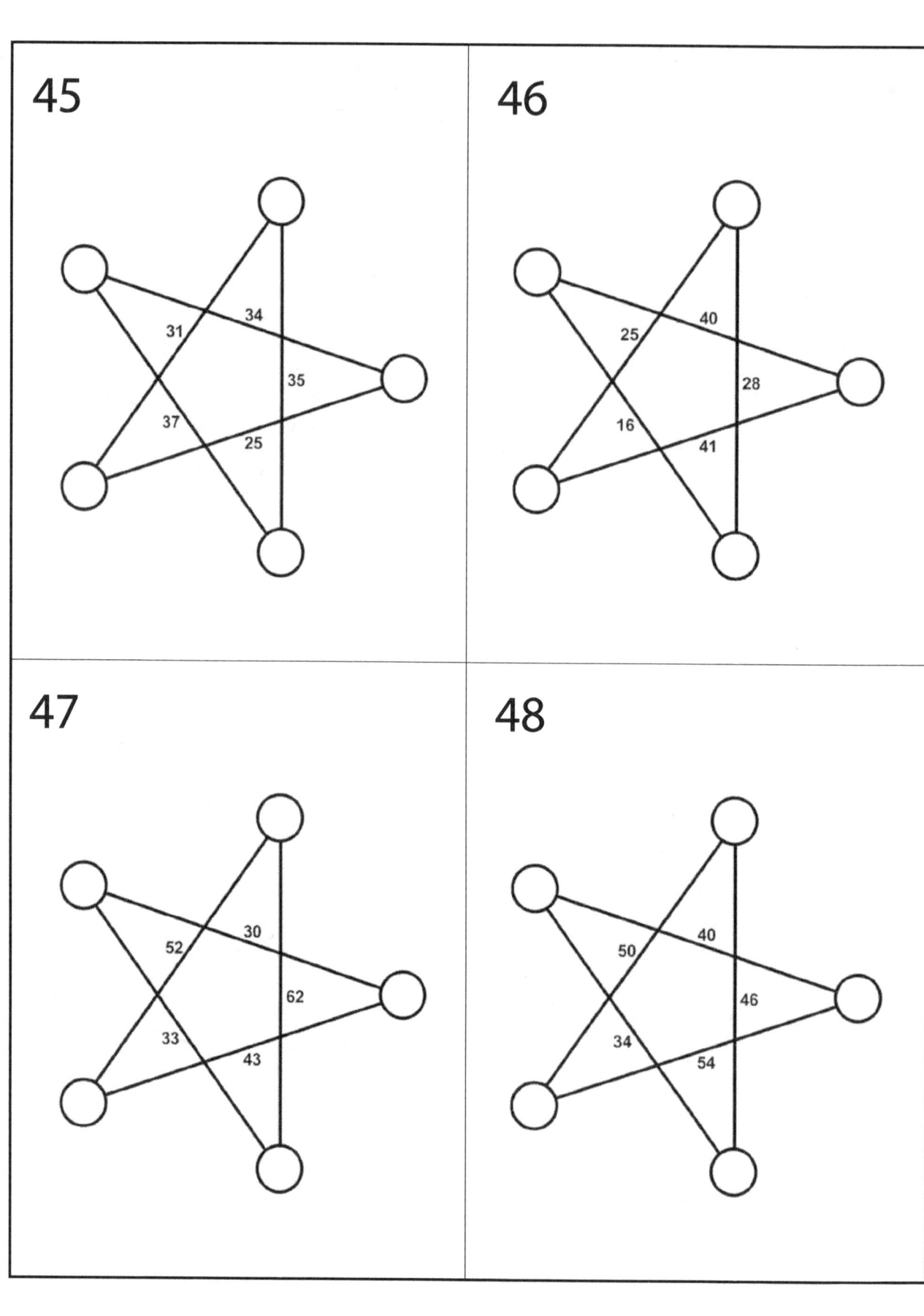

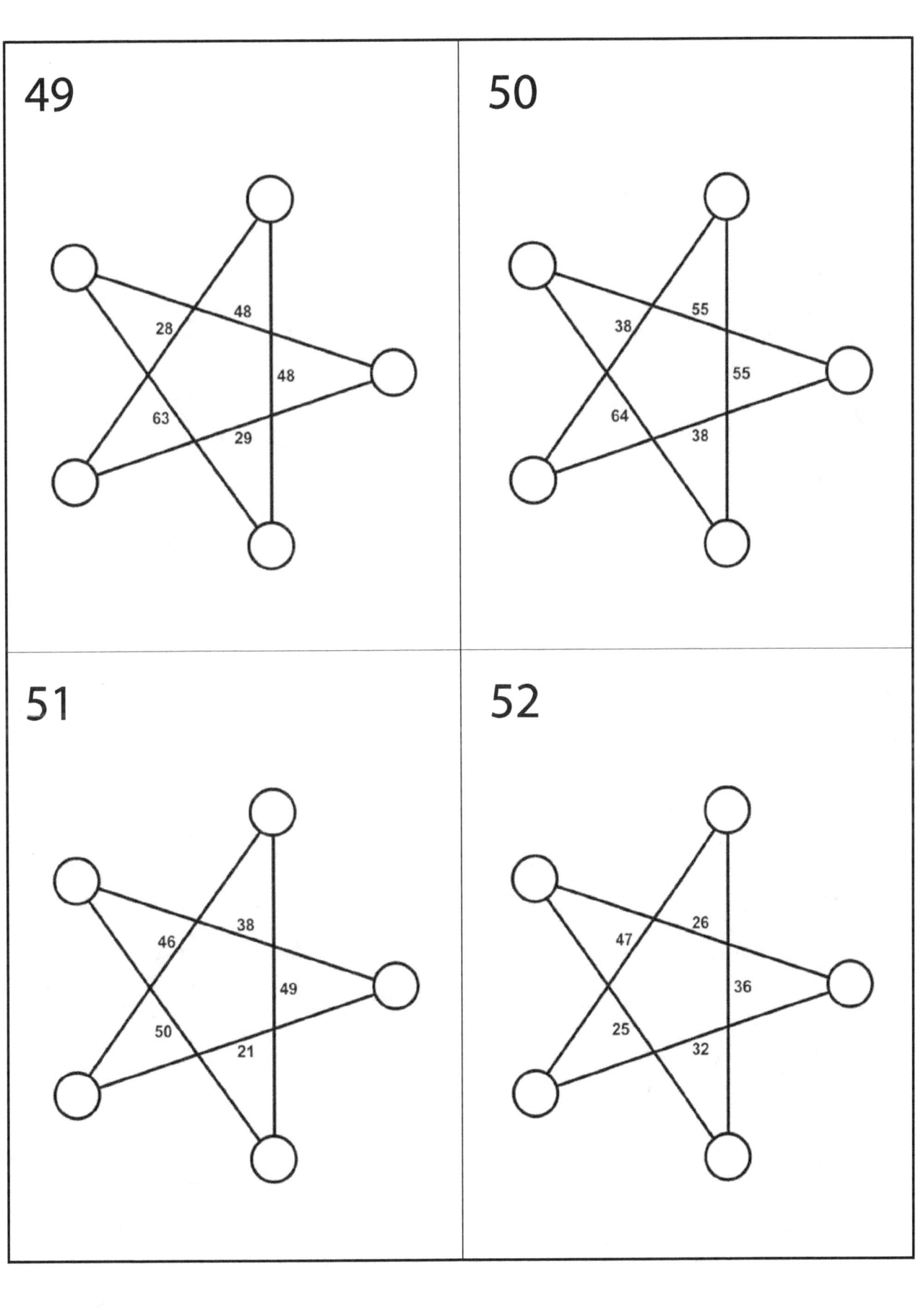

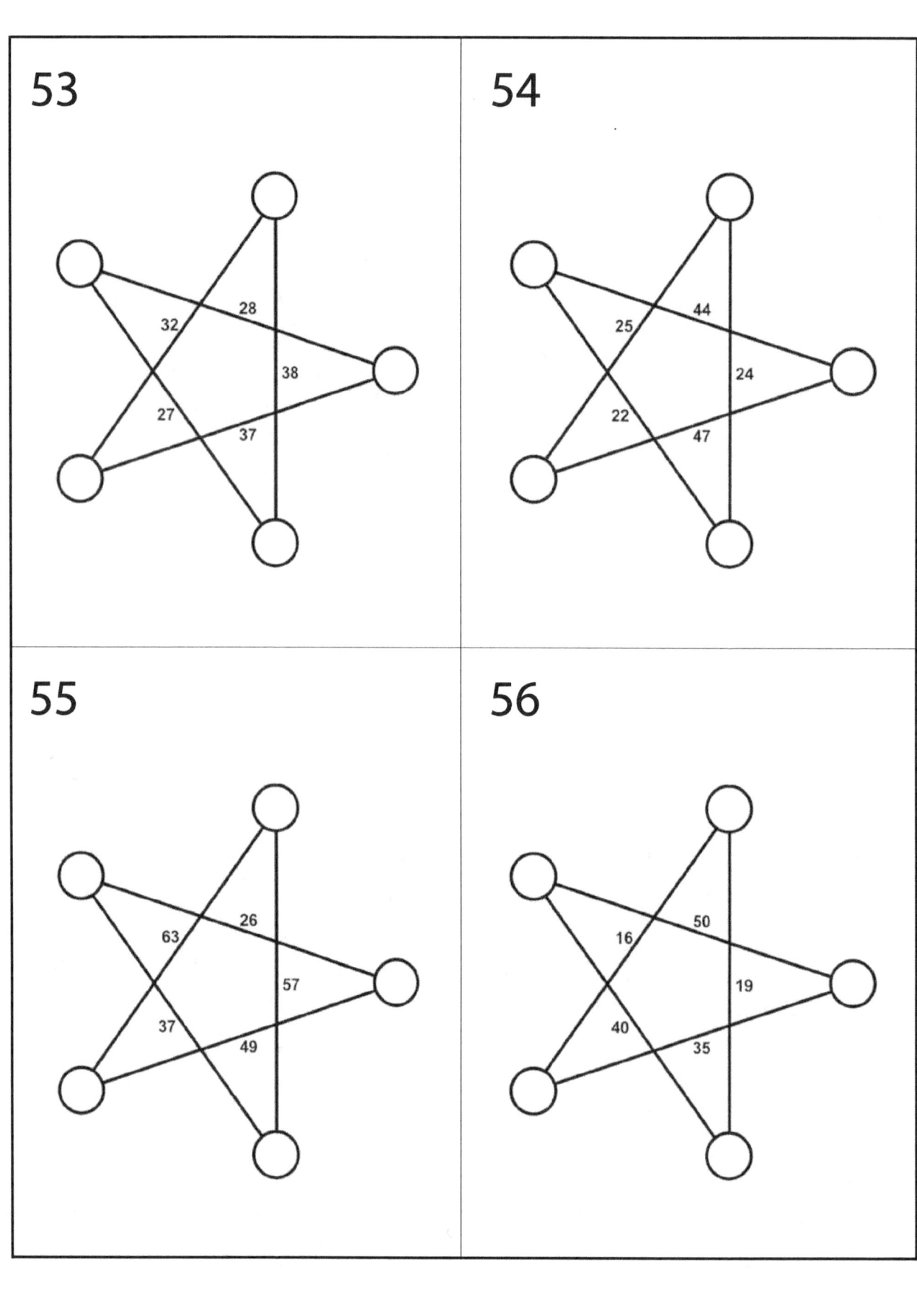

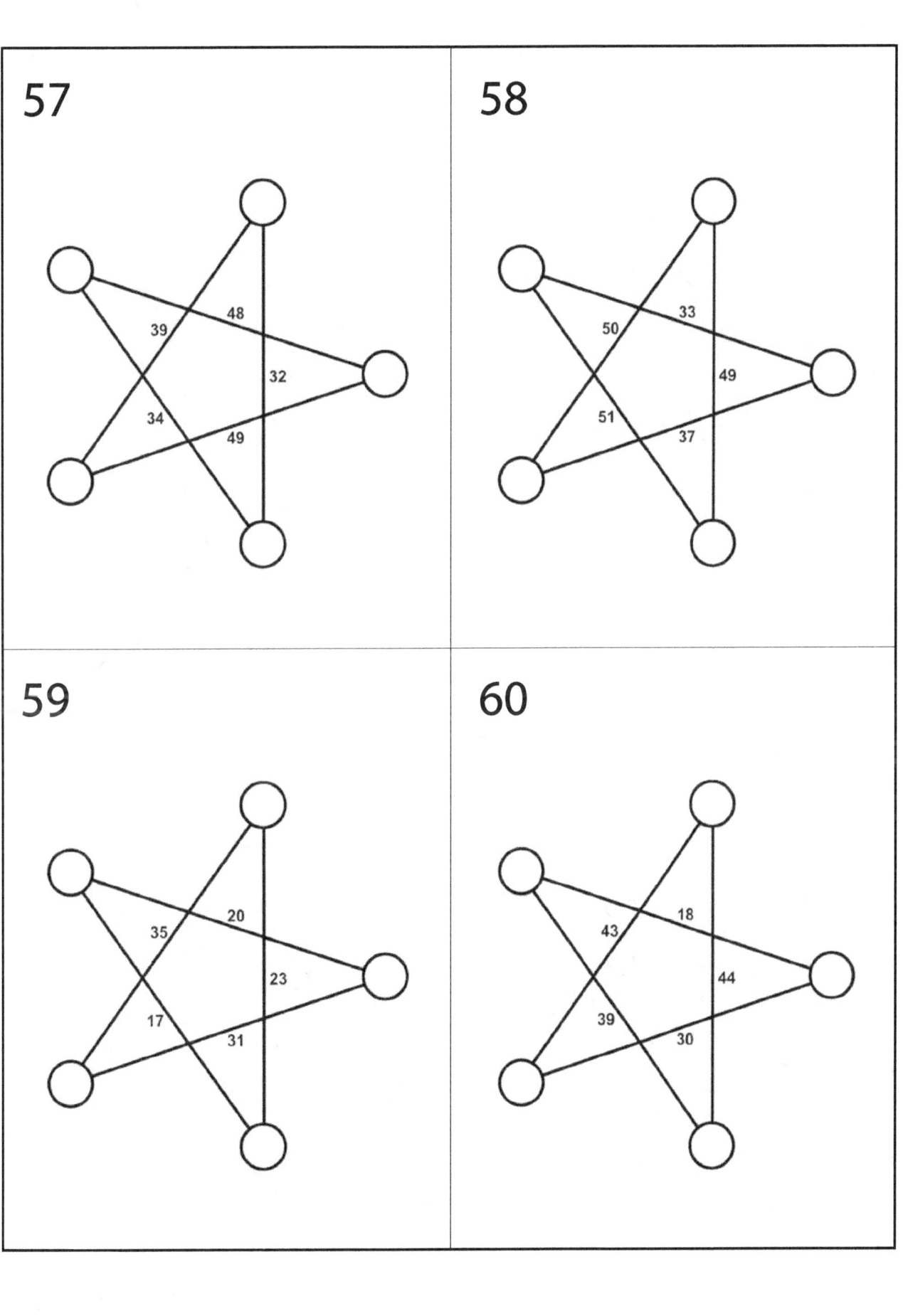

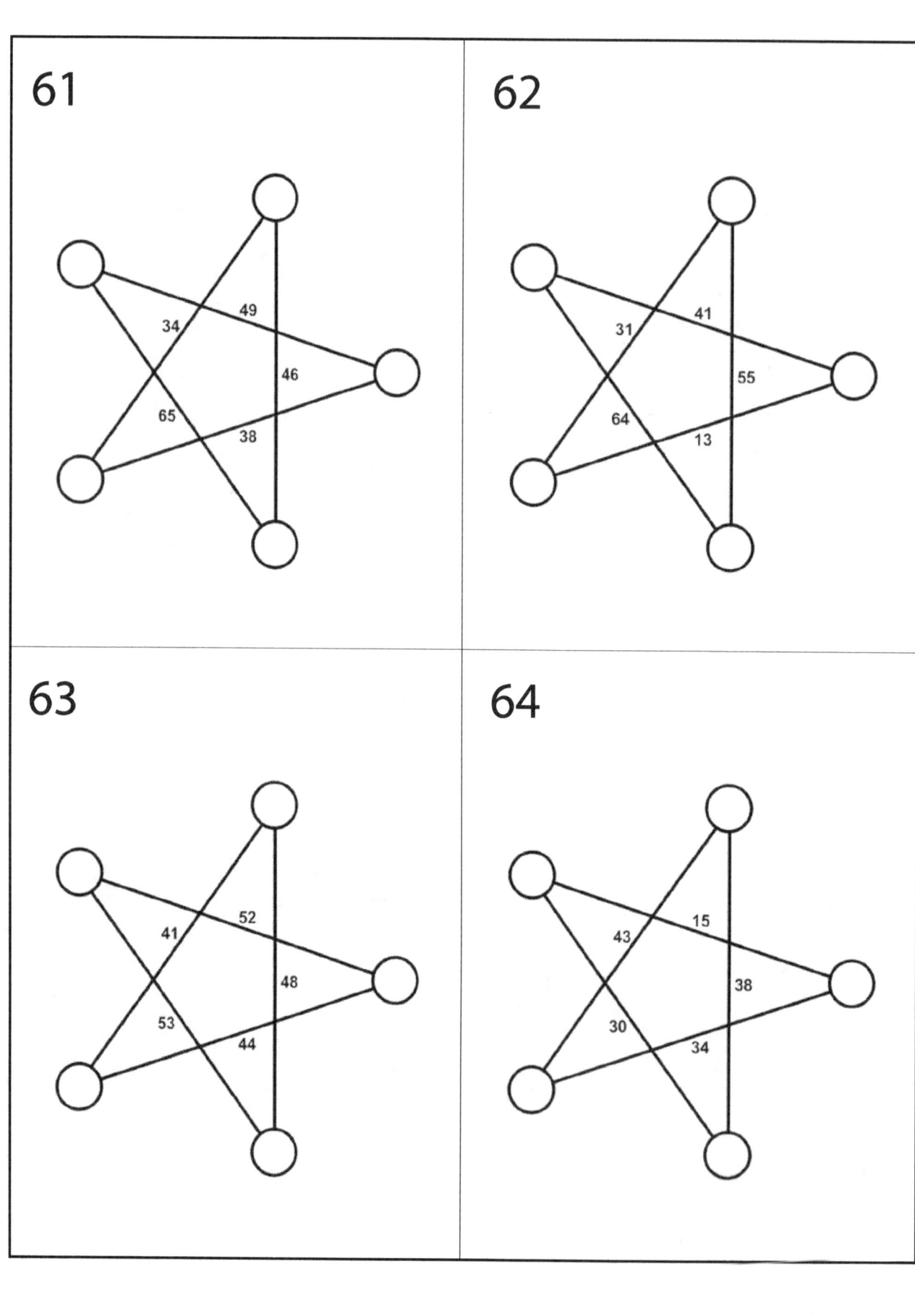

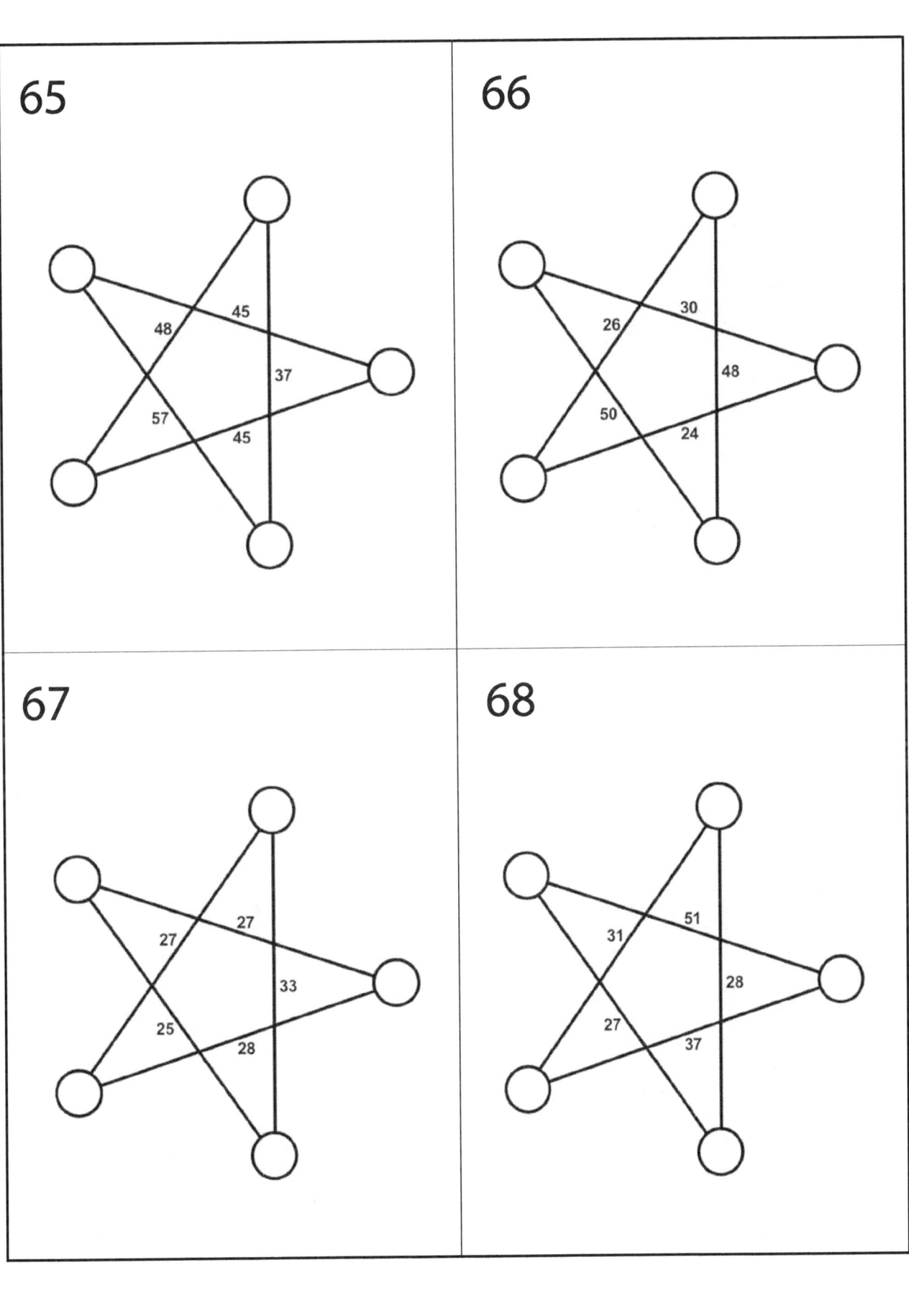

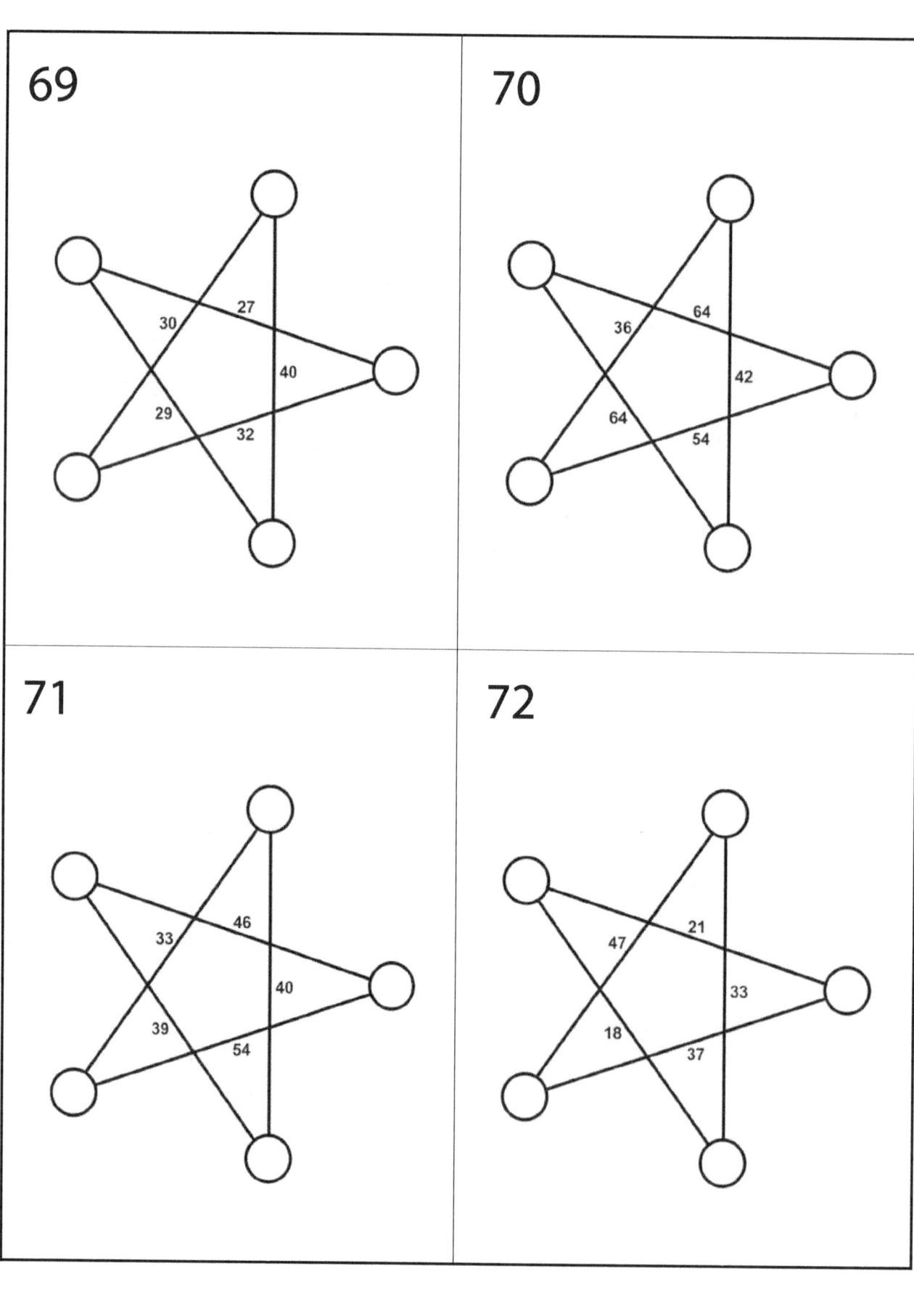

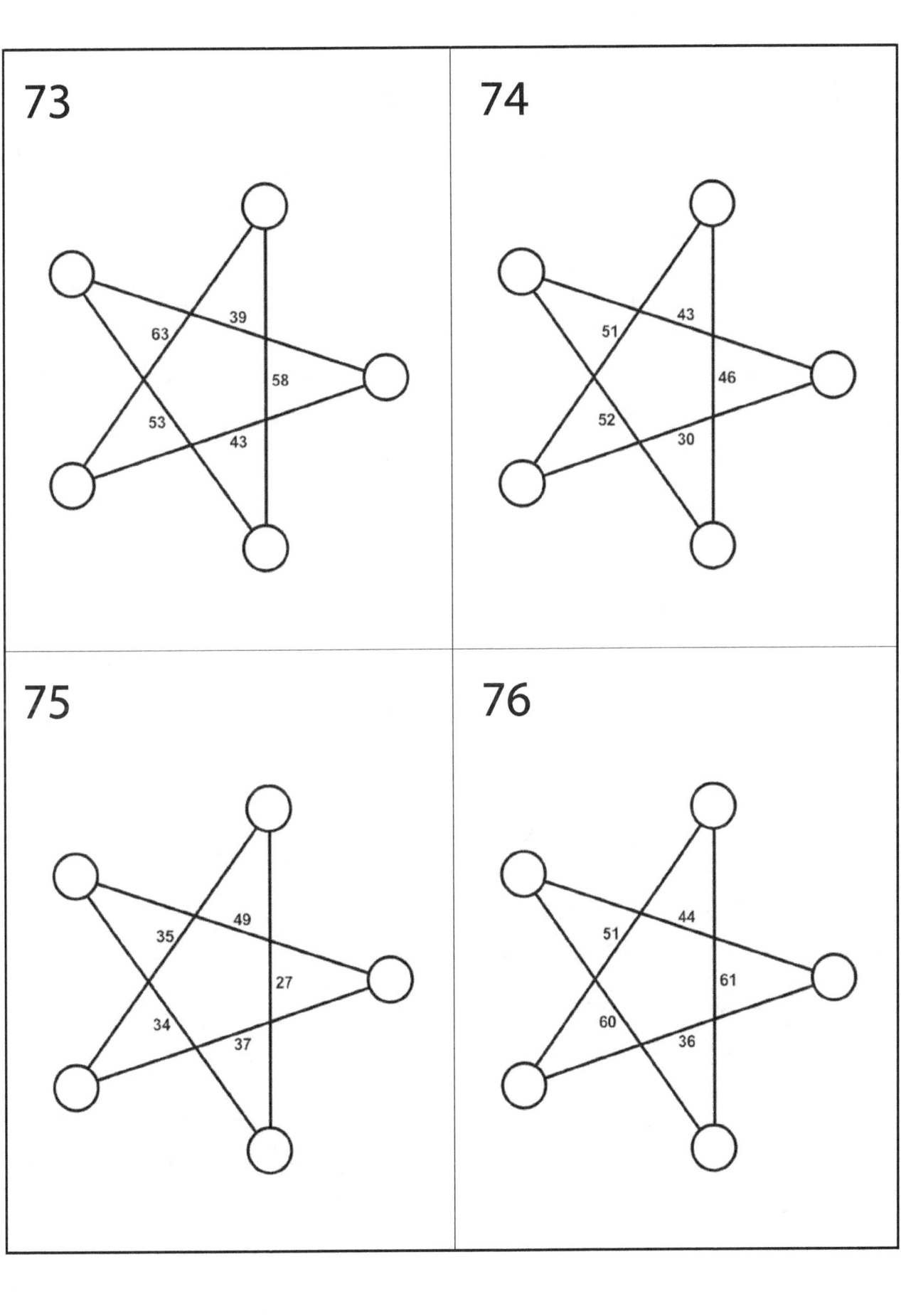

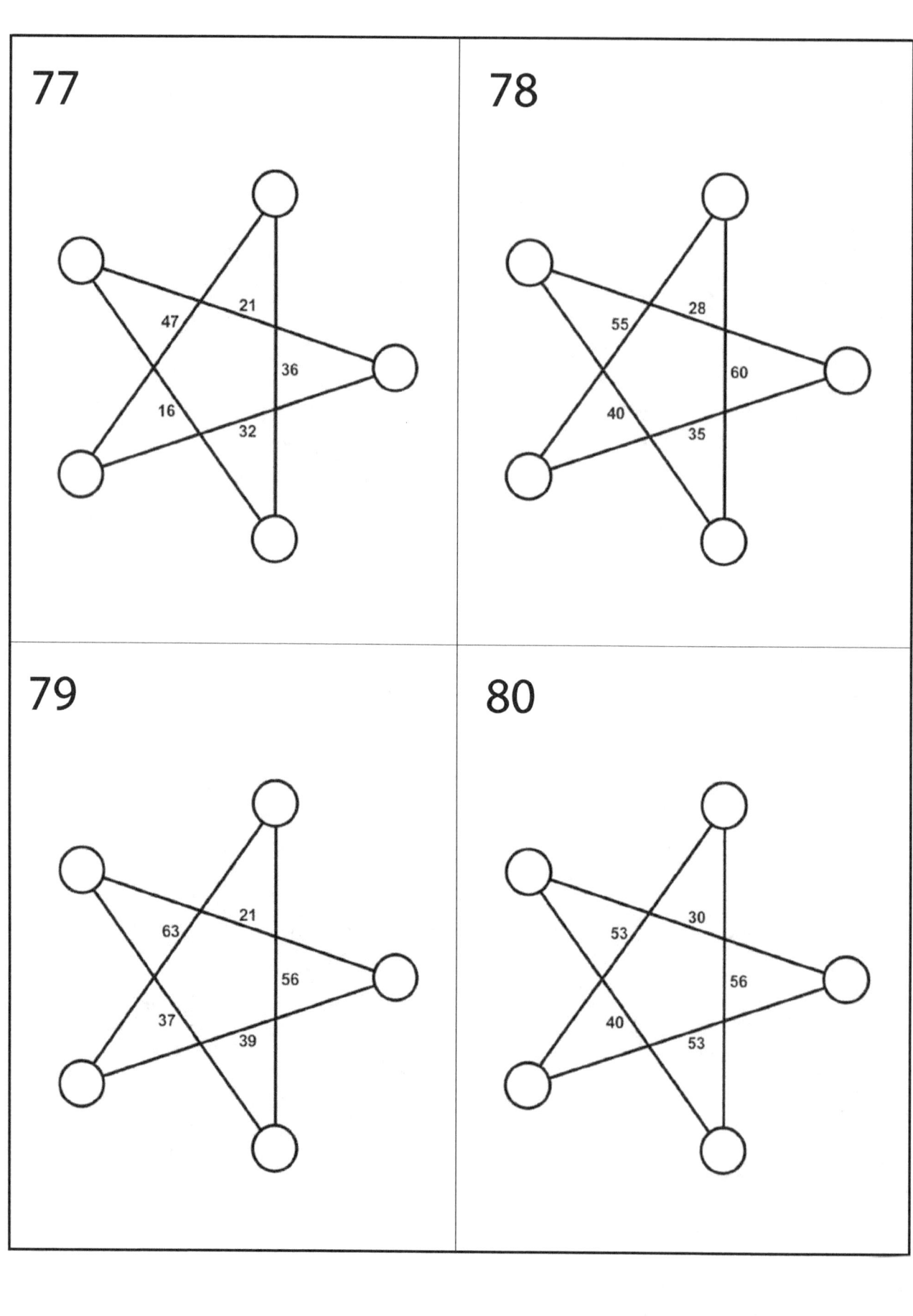

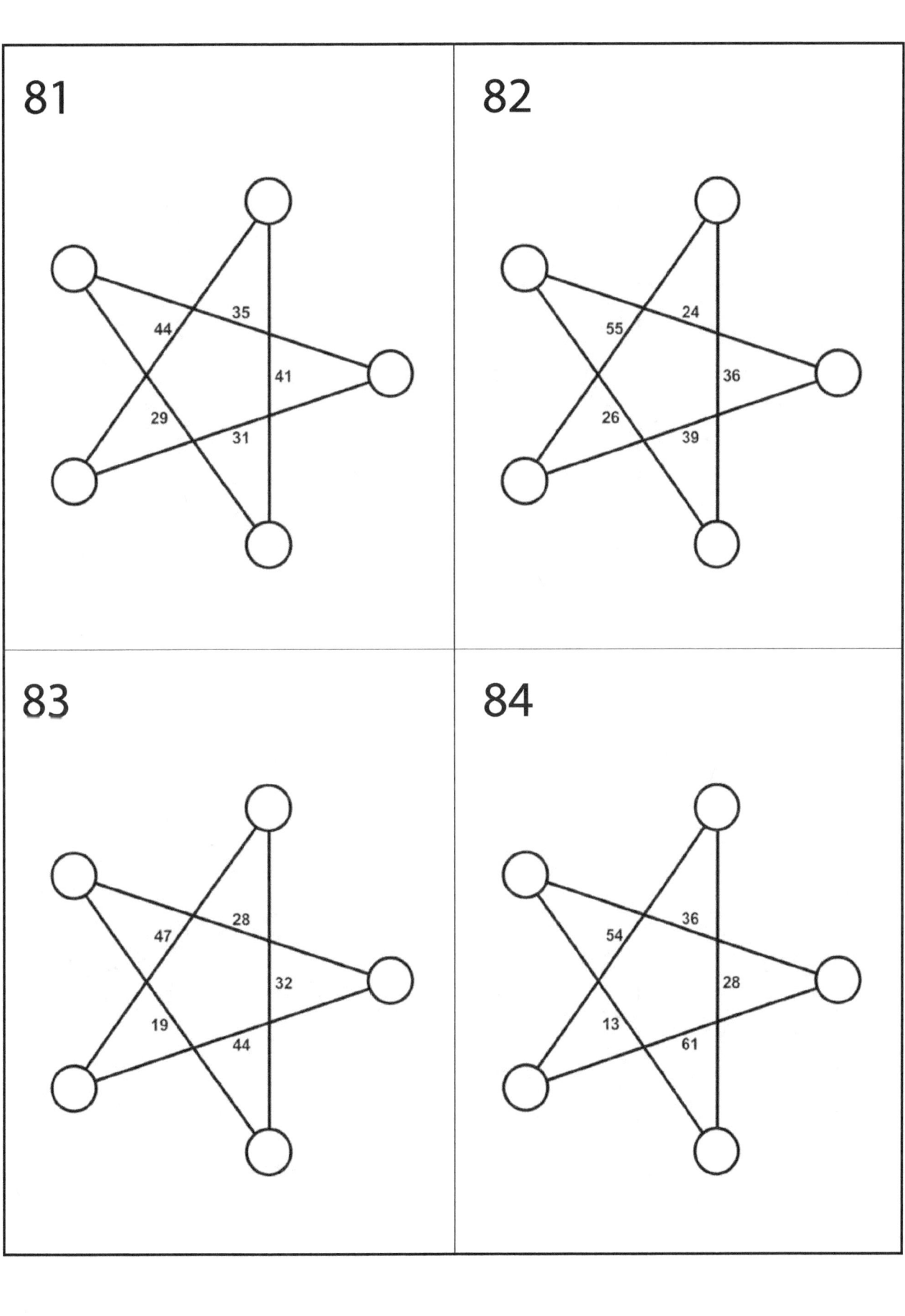

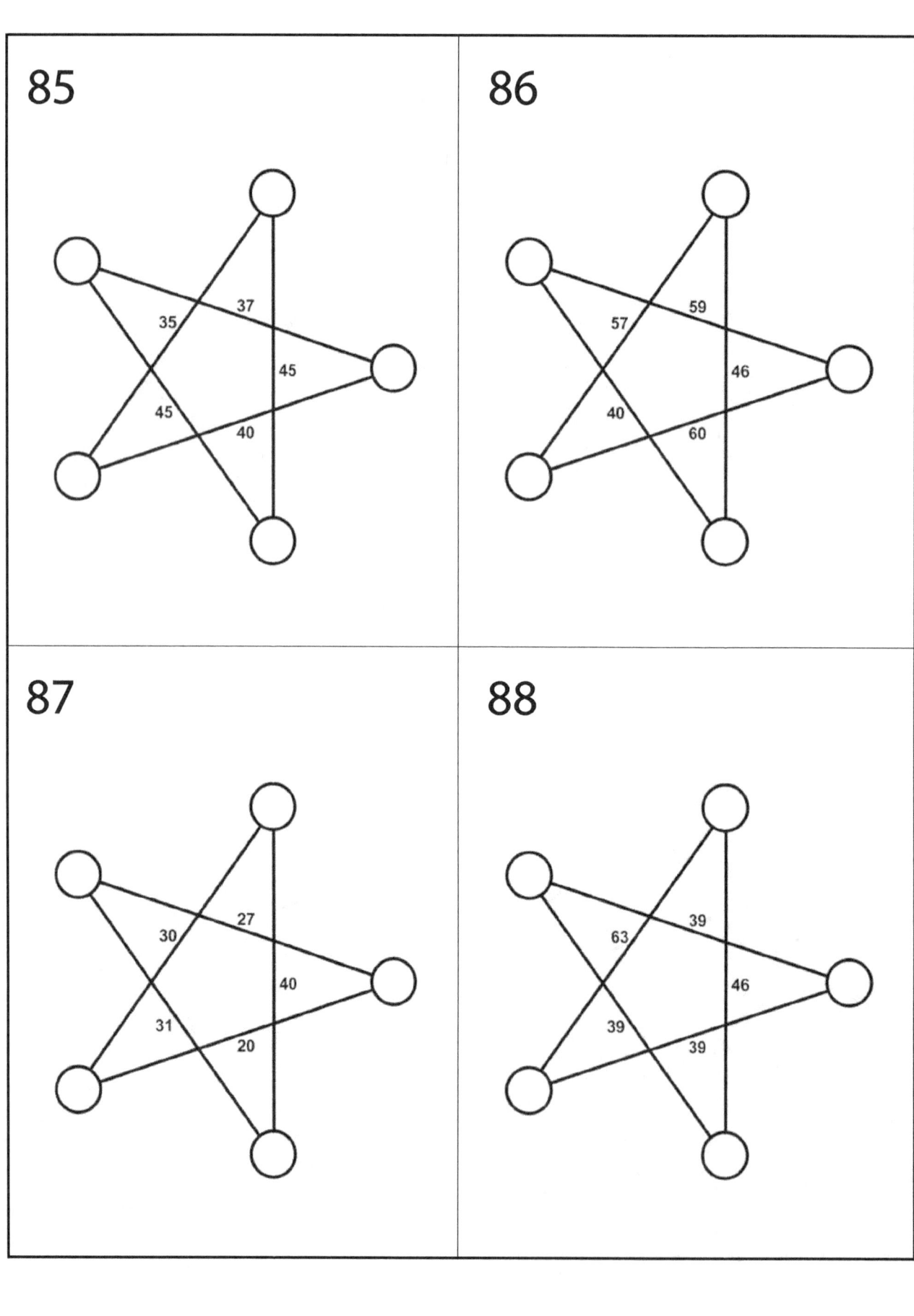

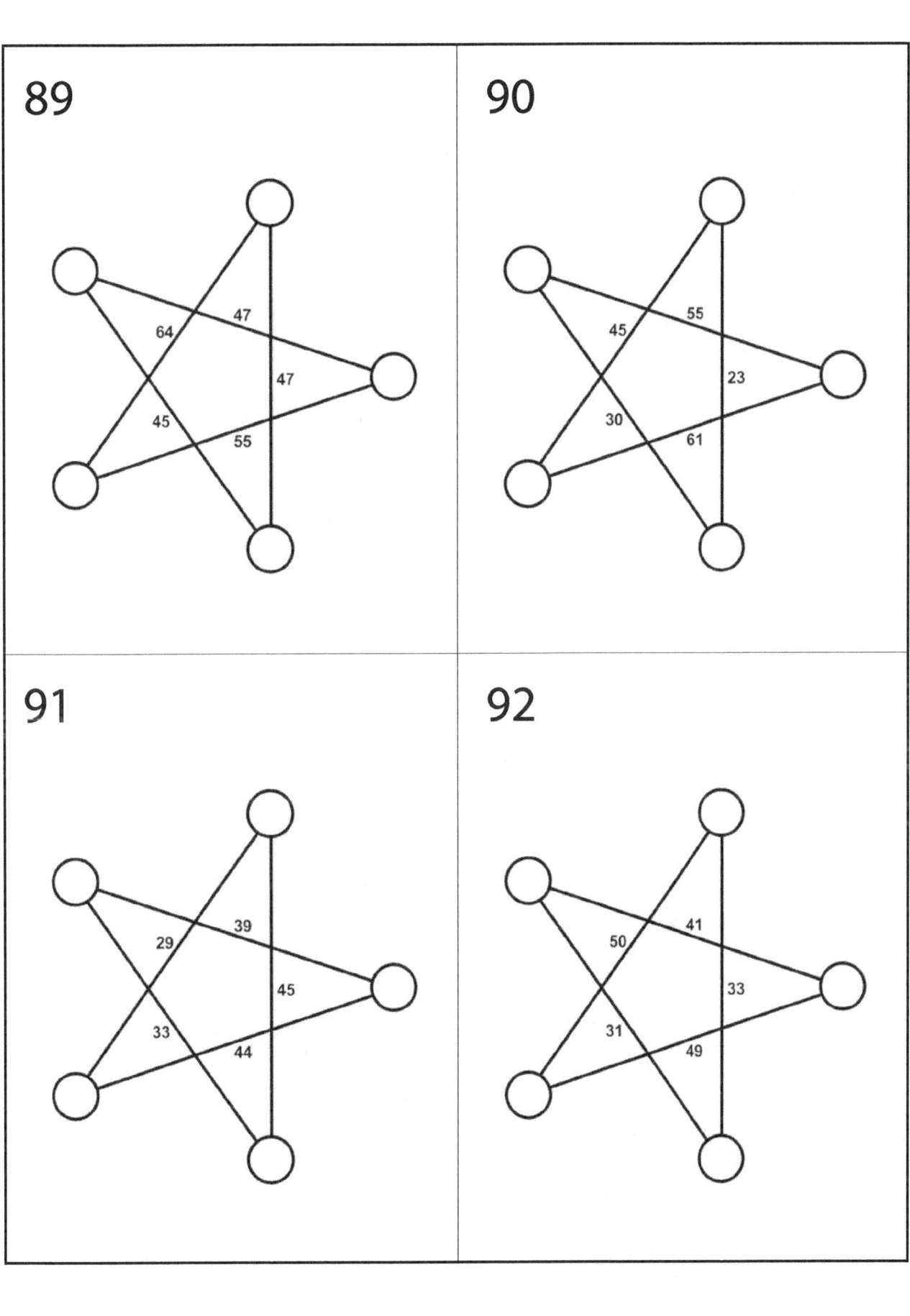

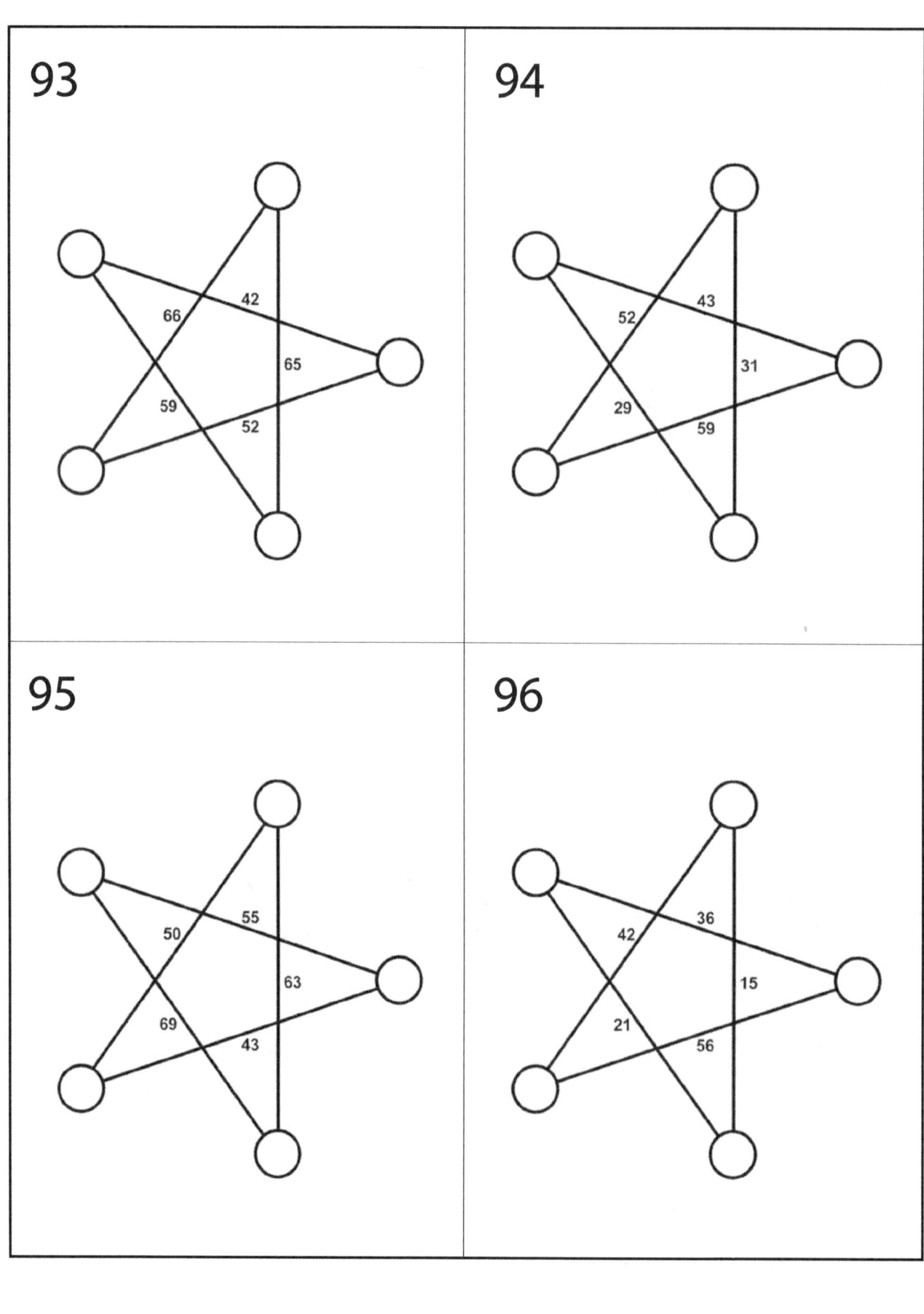

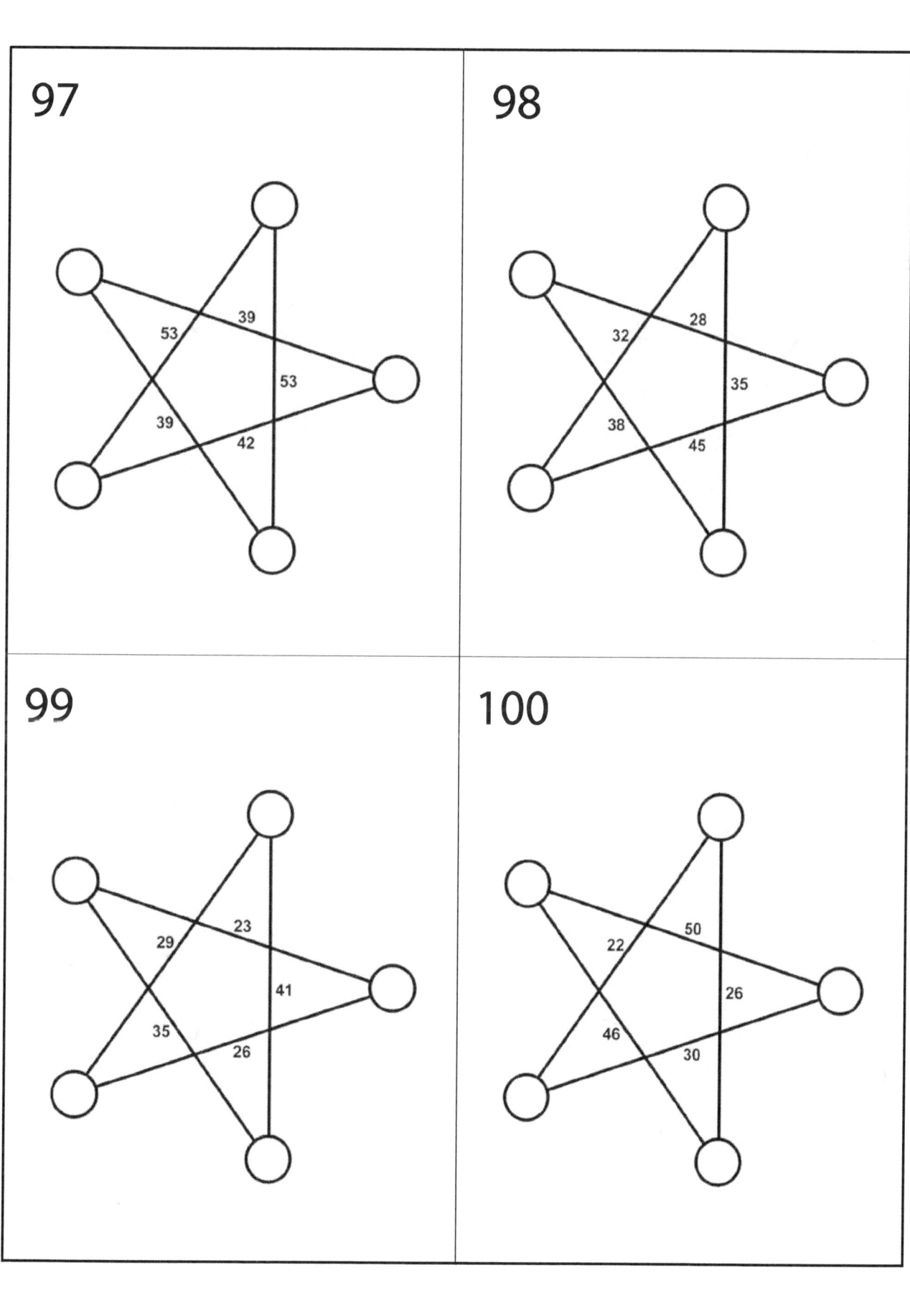

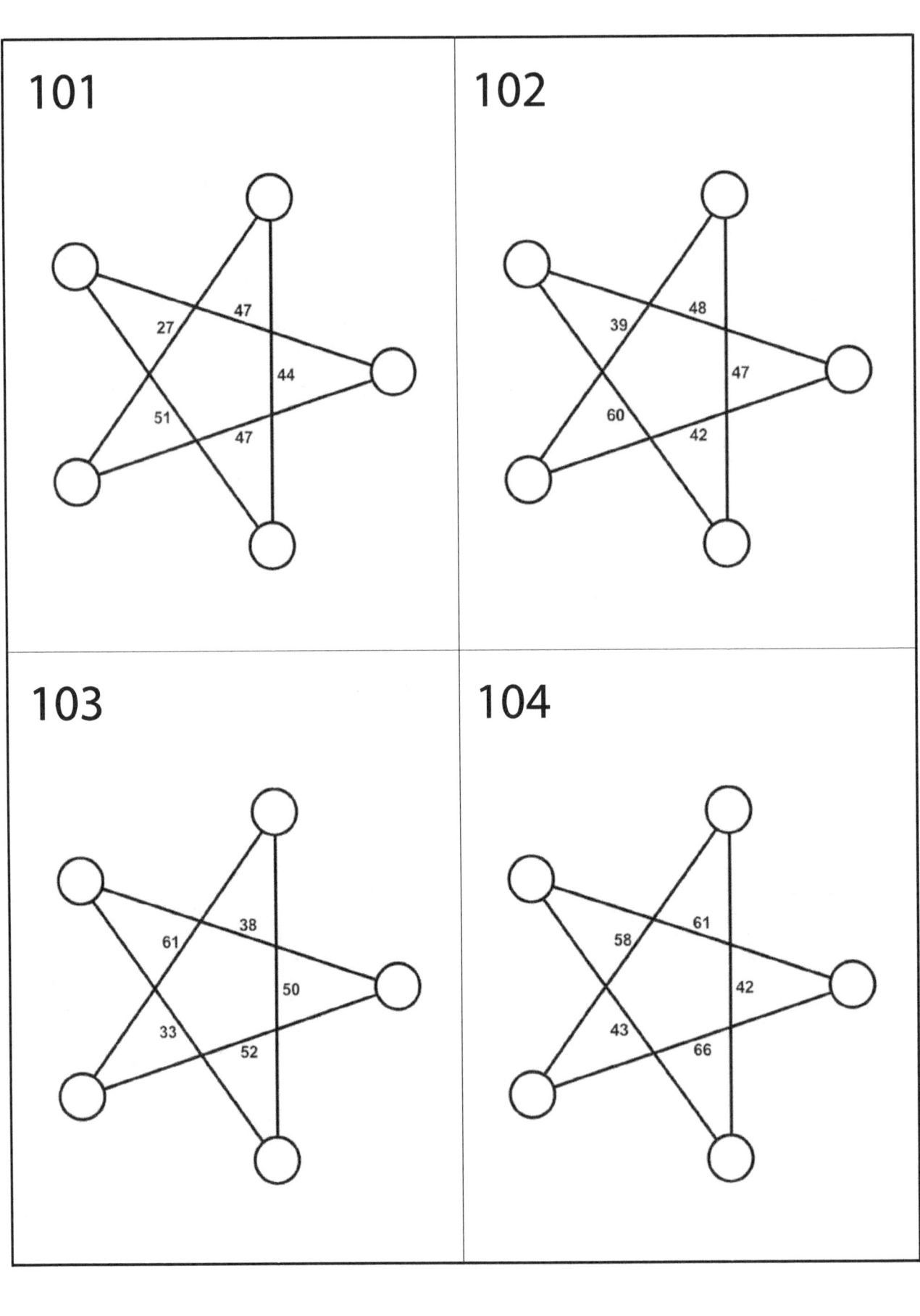

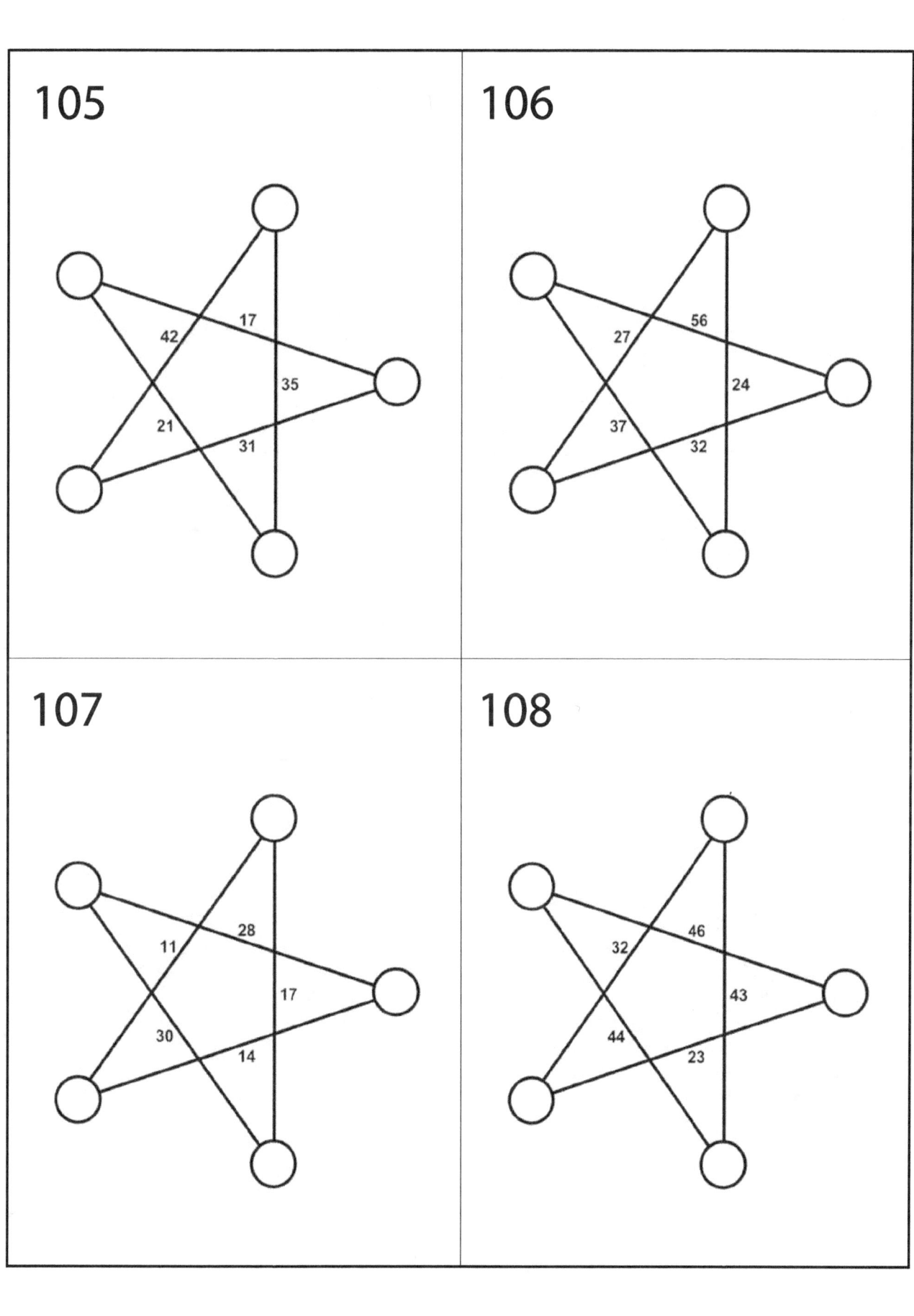

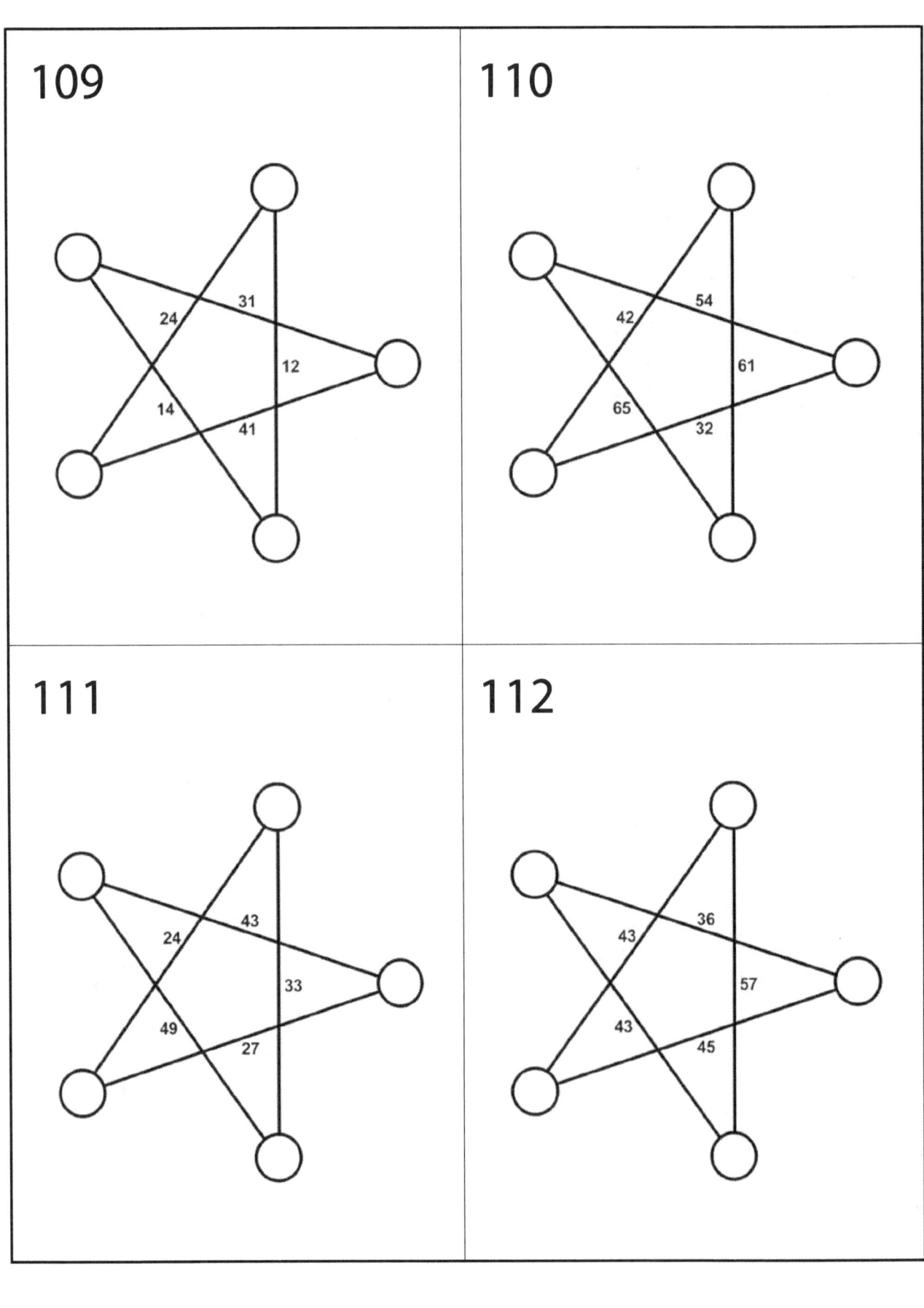

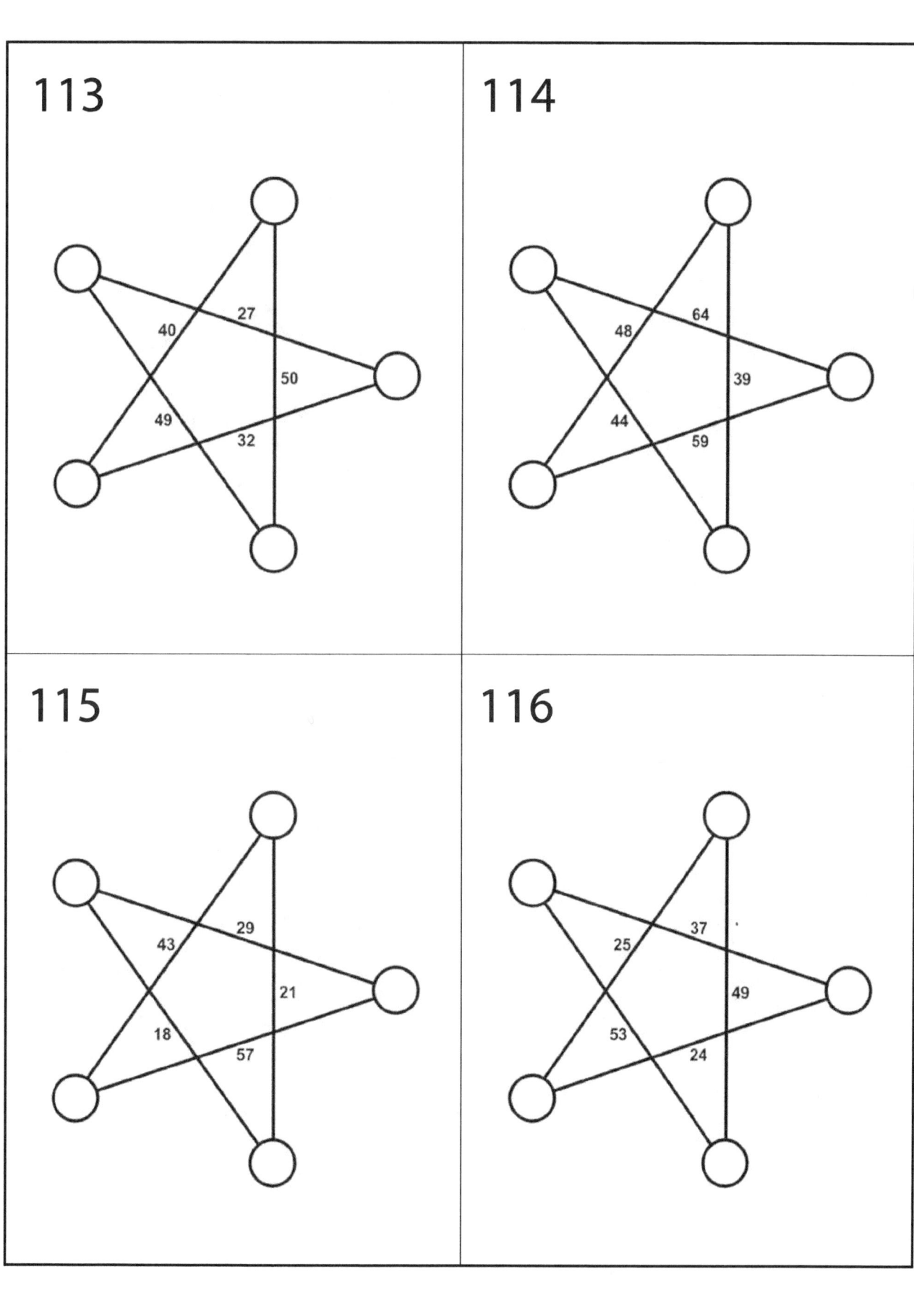

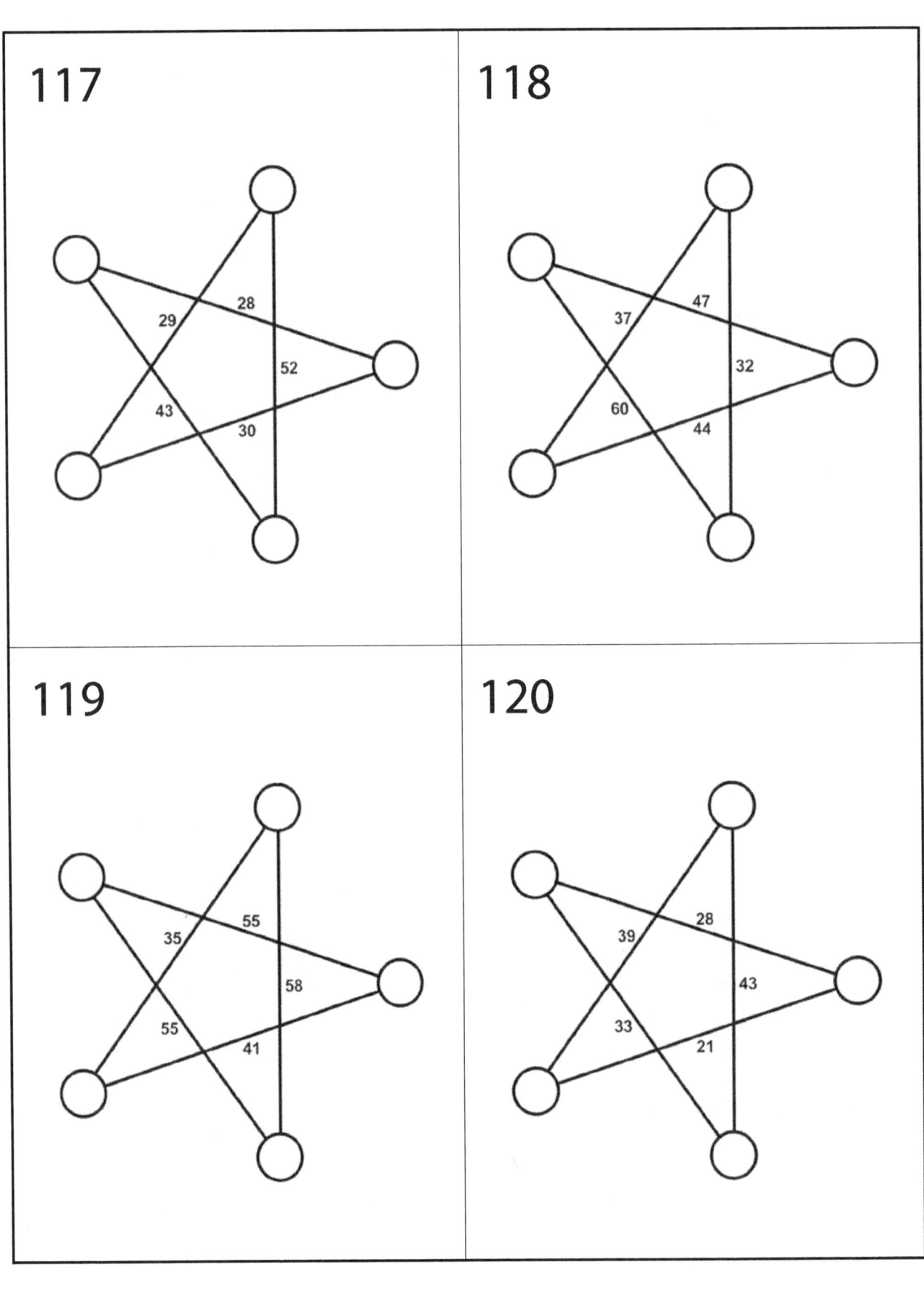

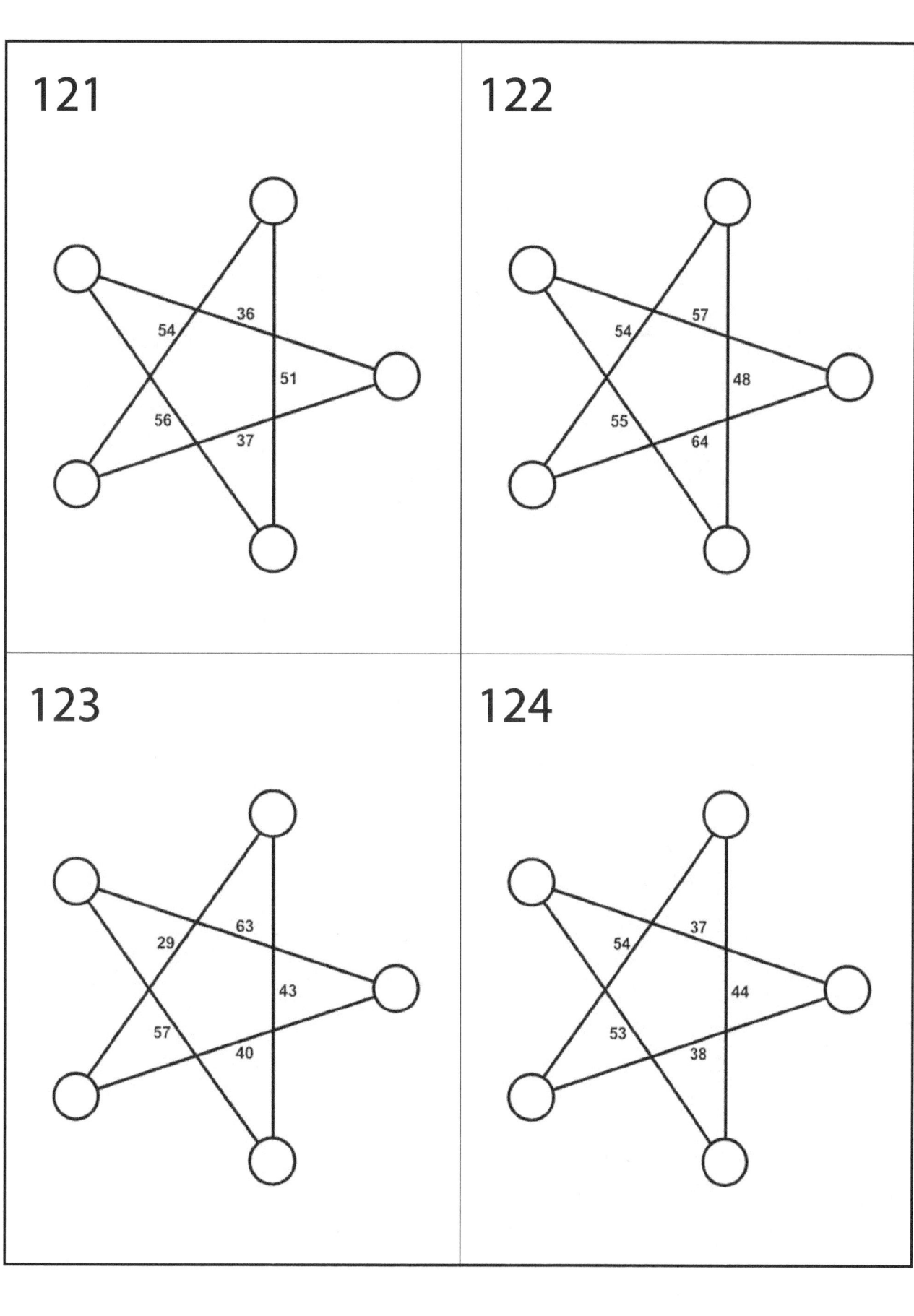

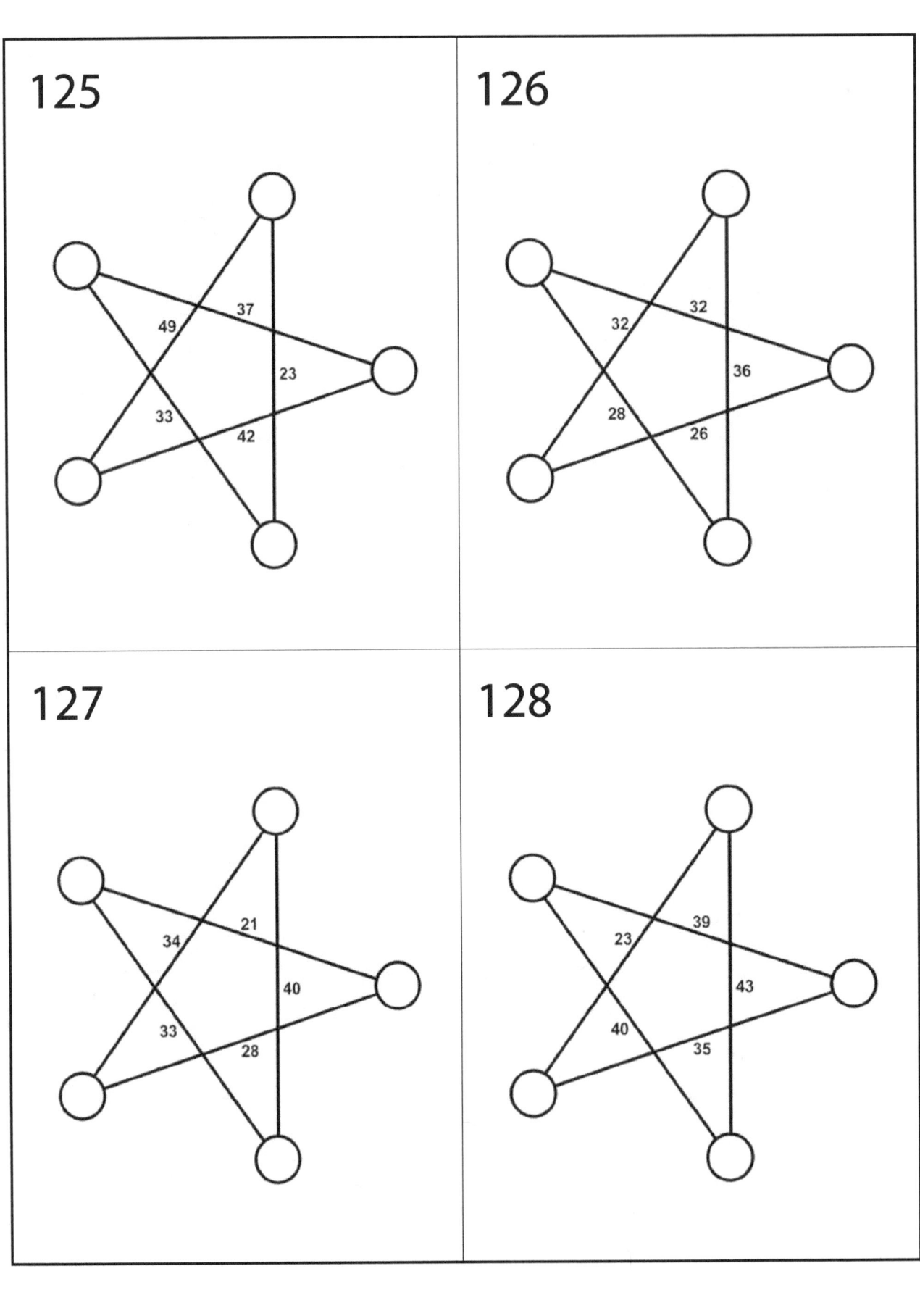

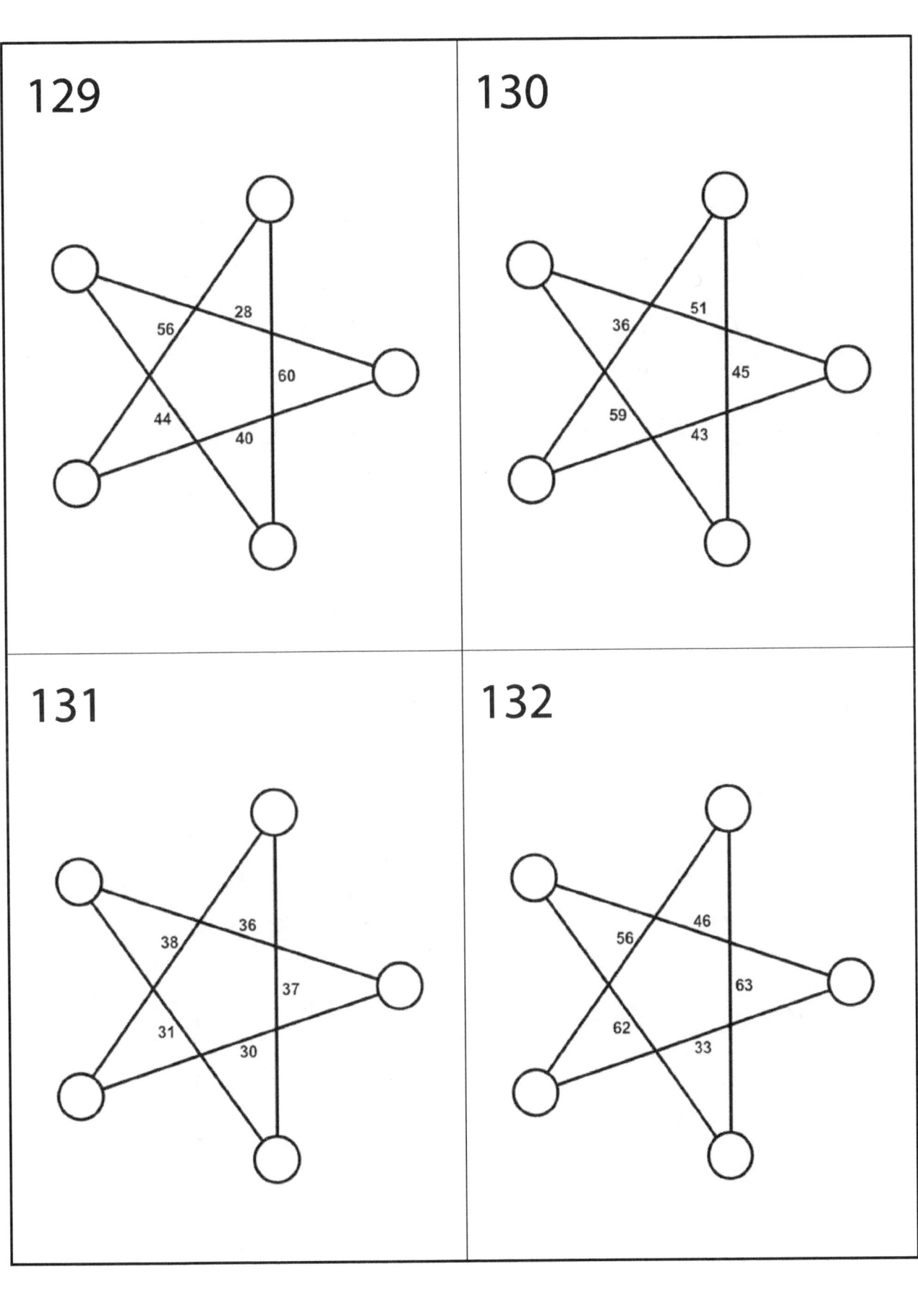

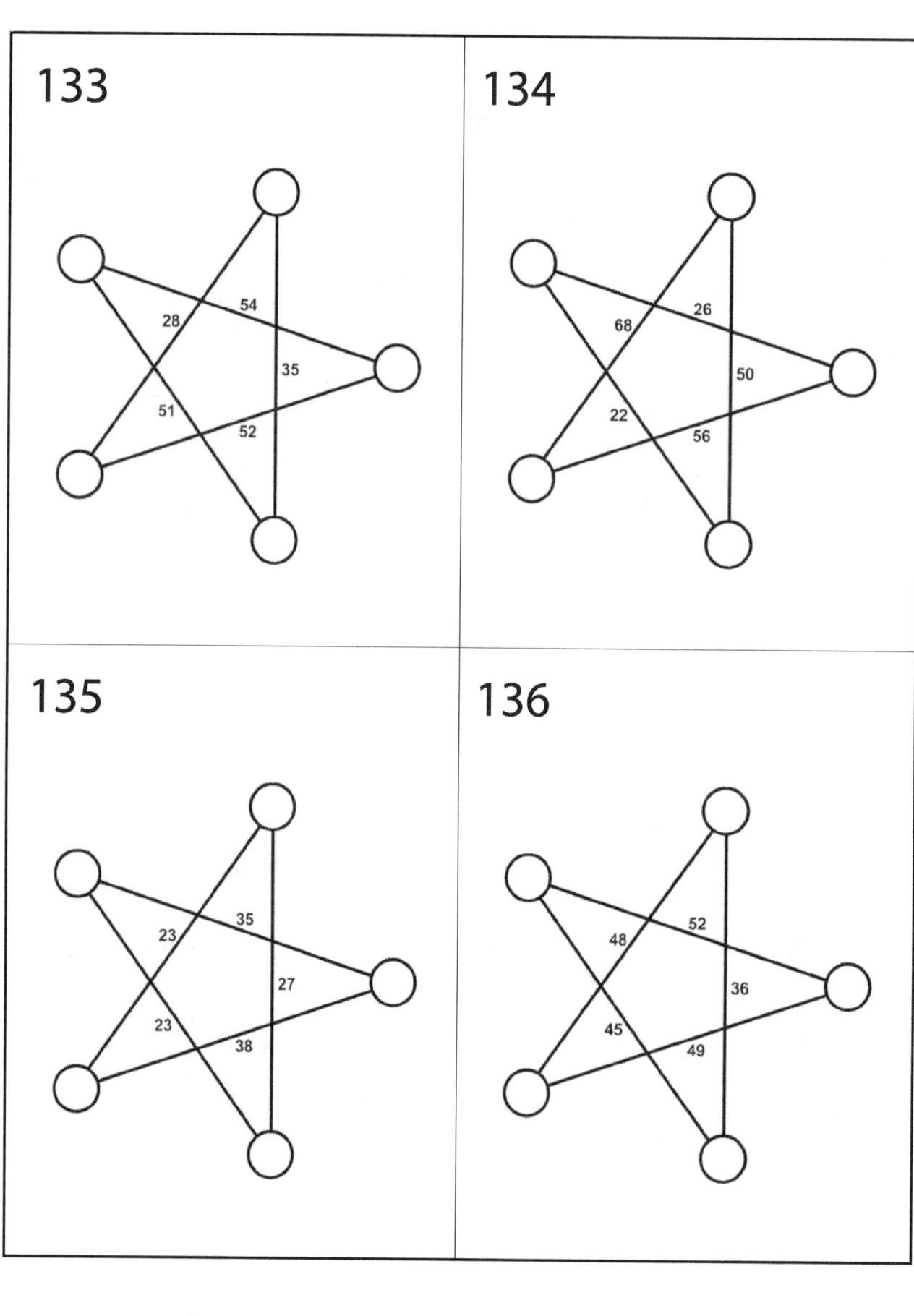

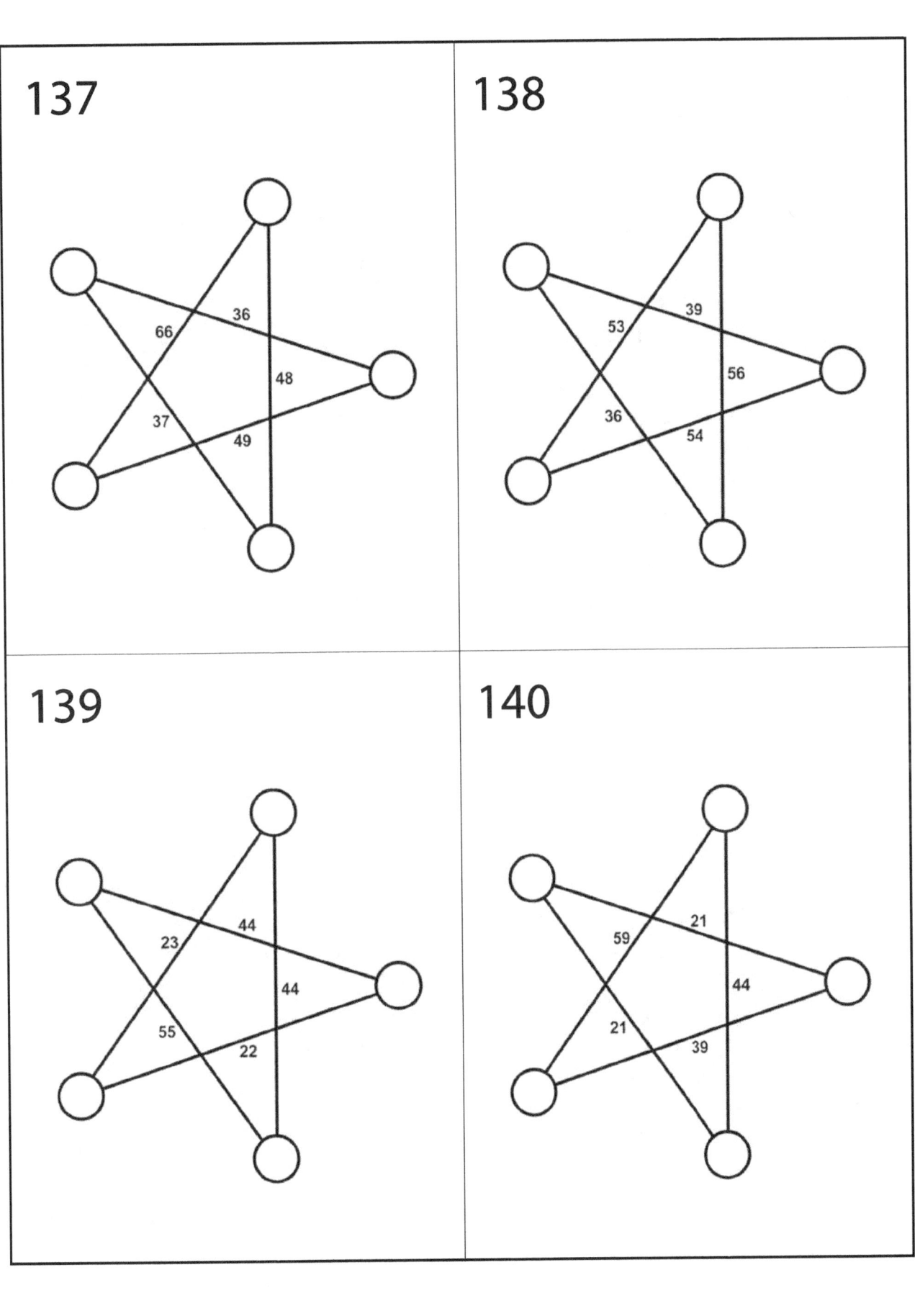

## 141

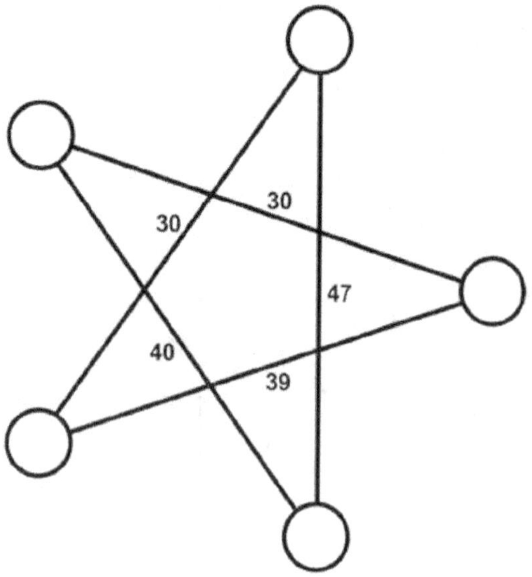

- 30
- 30
- 47
- 40
- 39

## 142

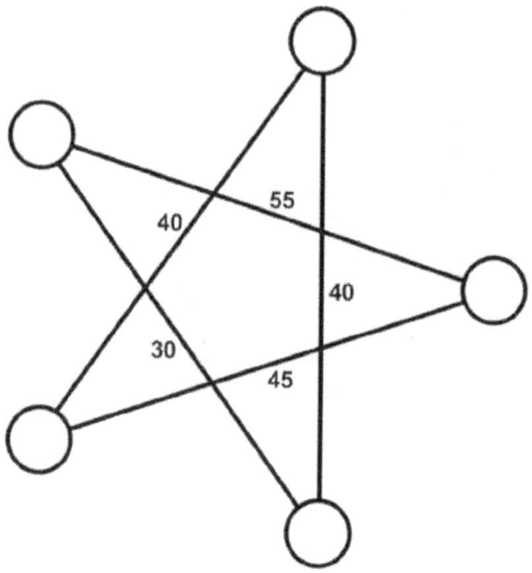

- 55
- 40
- 40
- 30
- 45

## 143

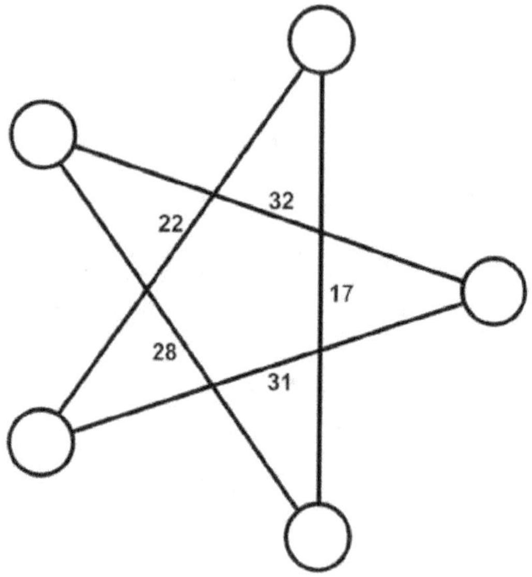

- 32
- 22
- 17
- 28
- 31

## 144

- 52
- 45
- 58
- 59
- 30

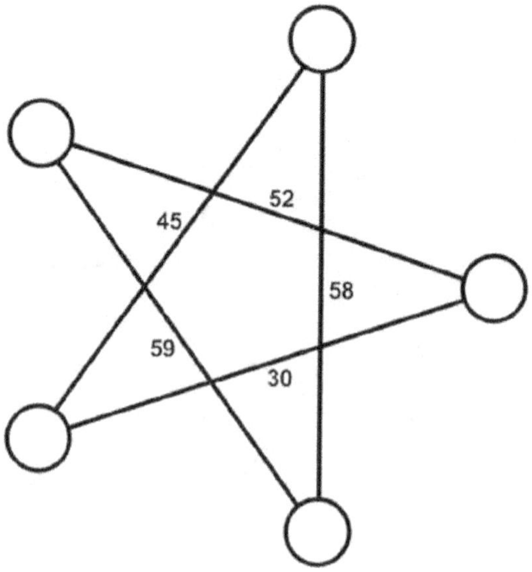

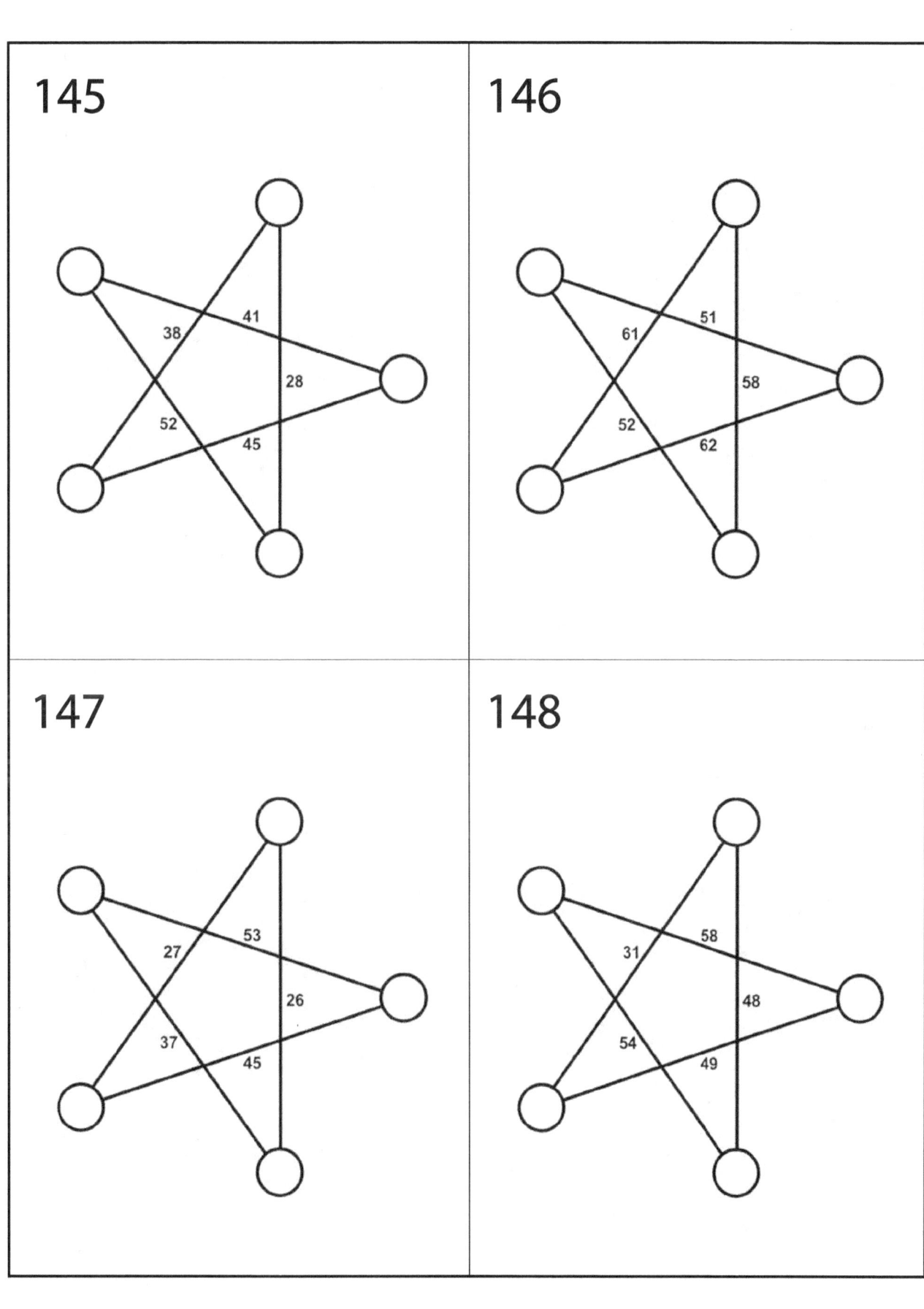

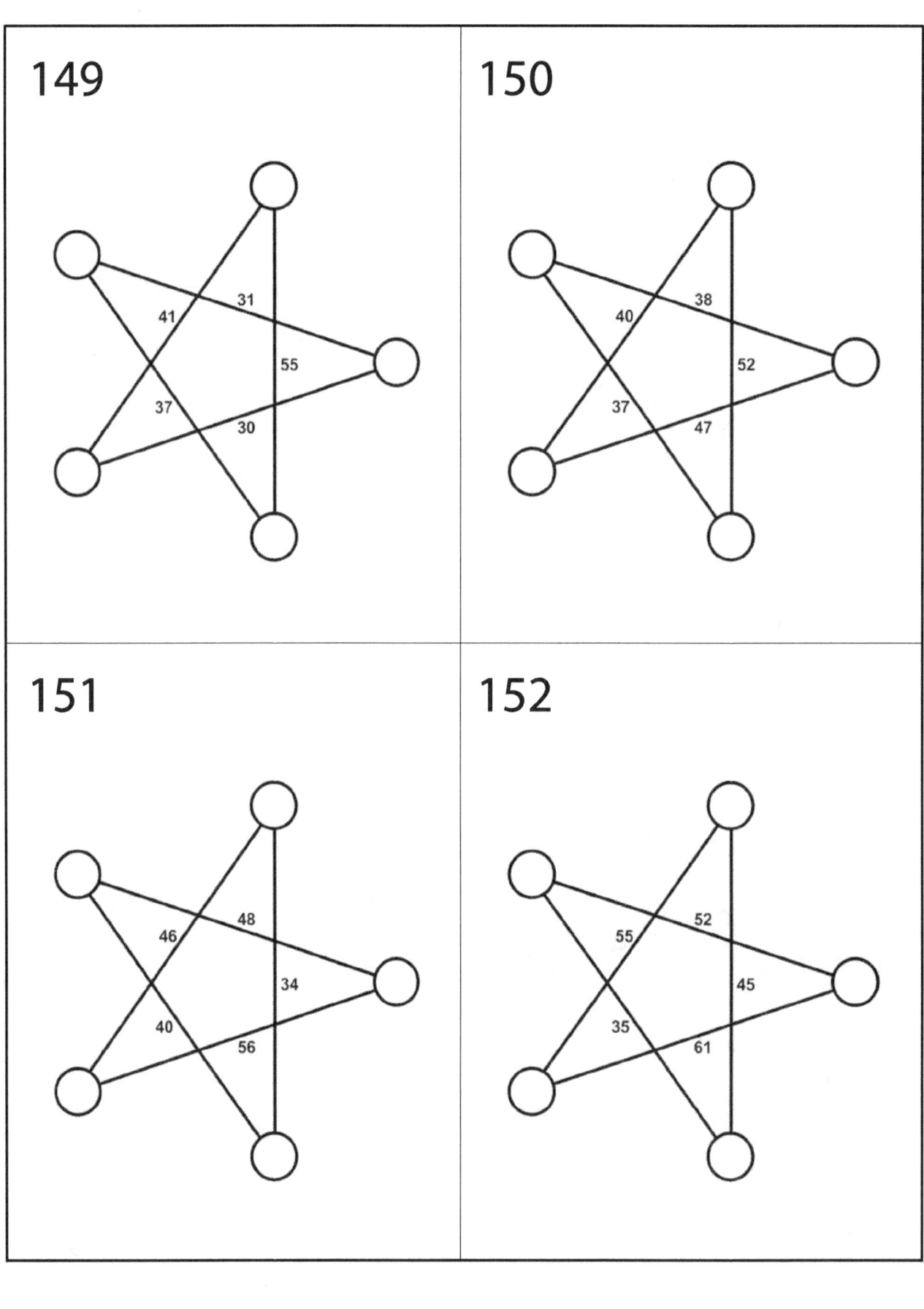

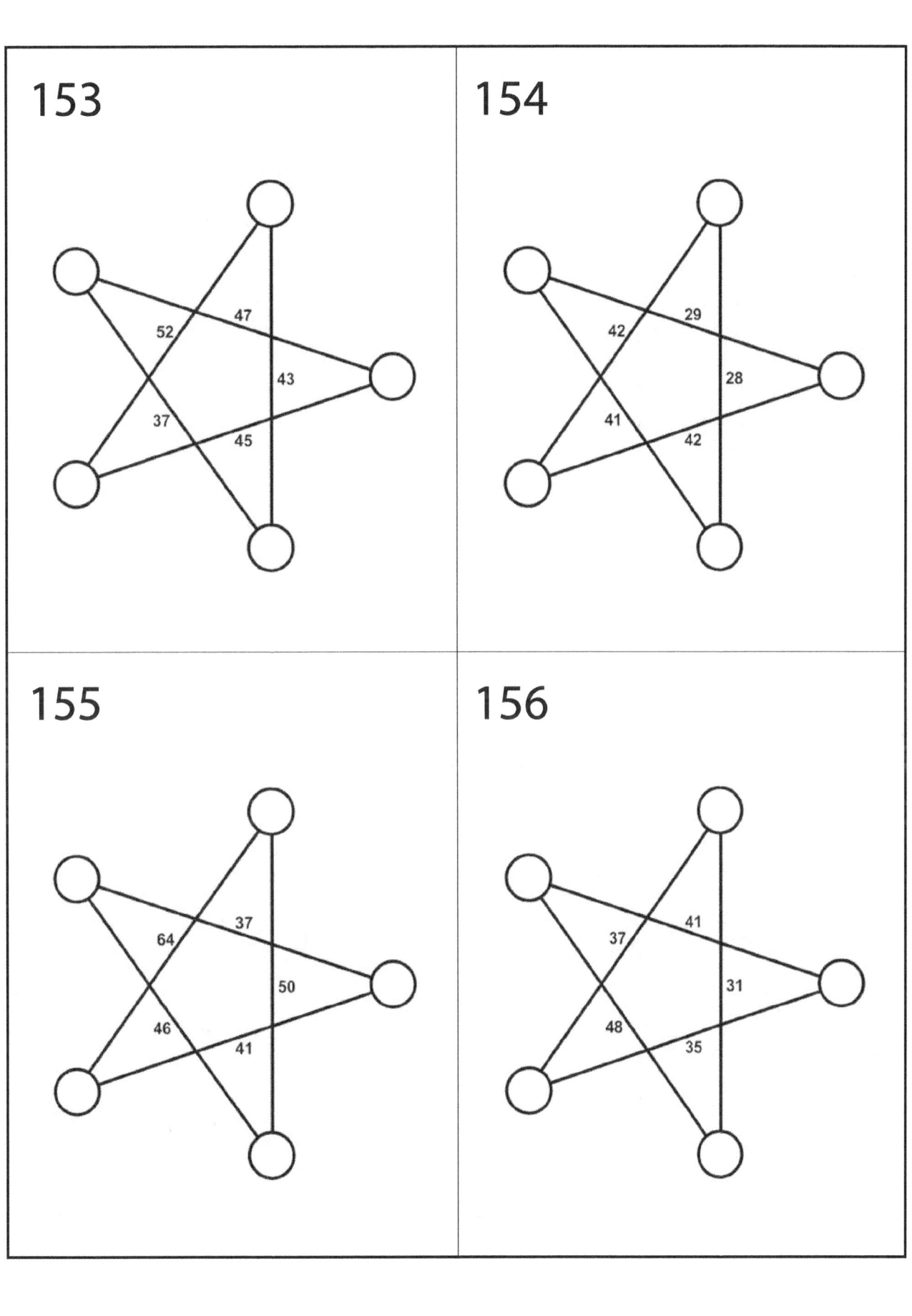

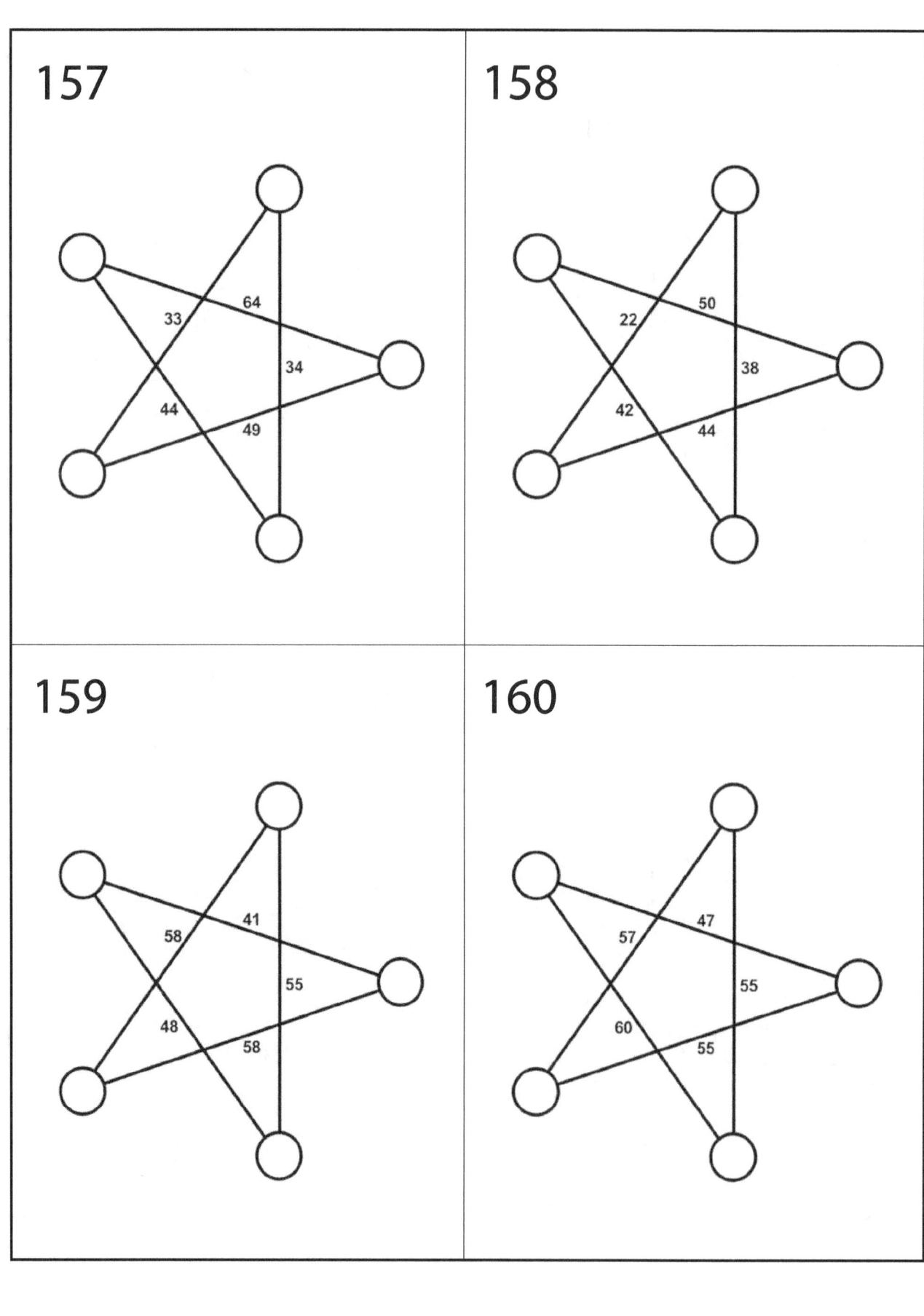

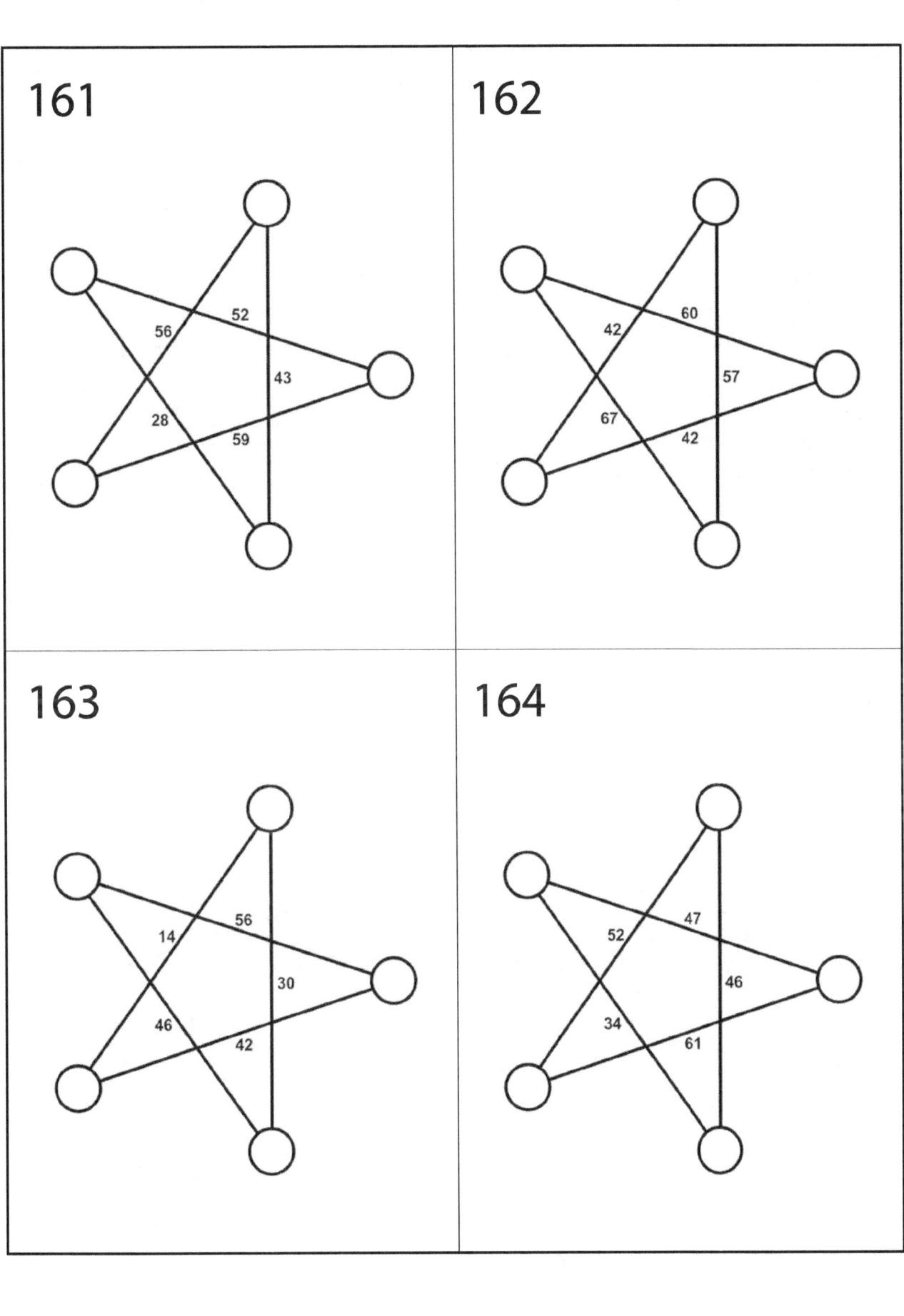

## 165

## 166

## 167

## 168

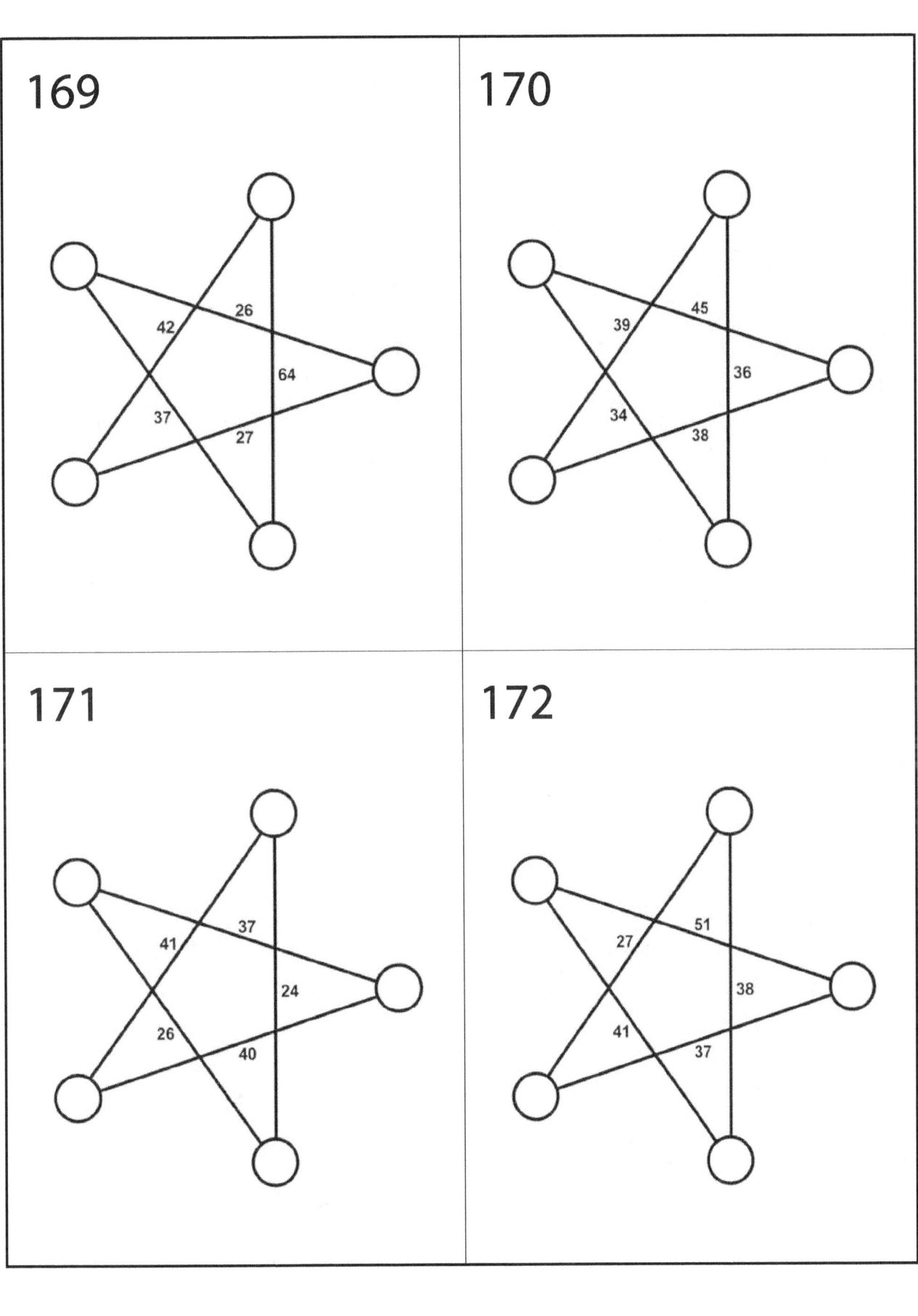

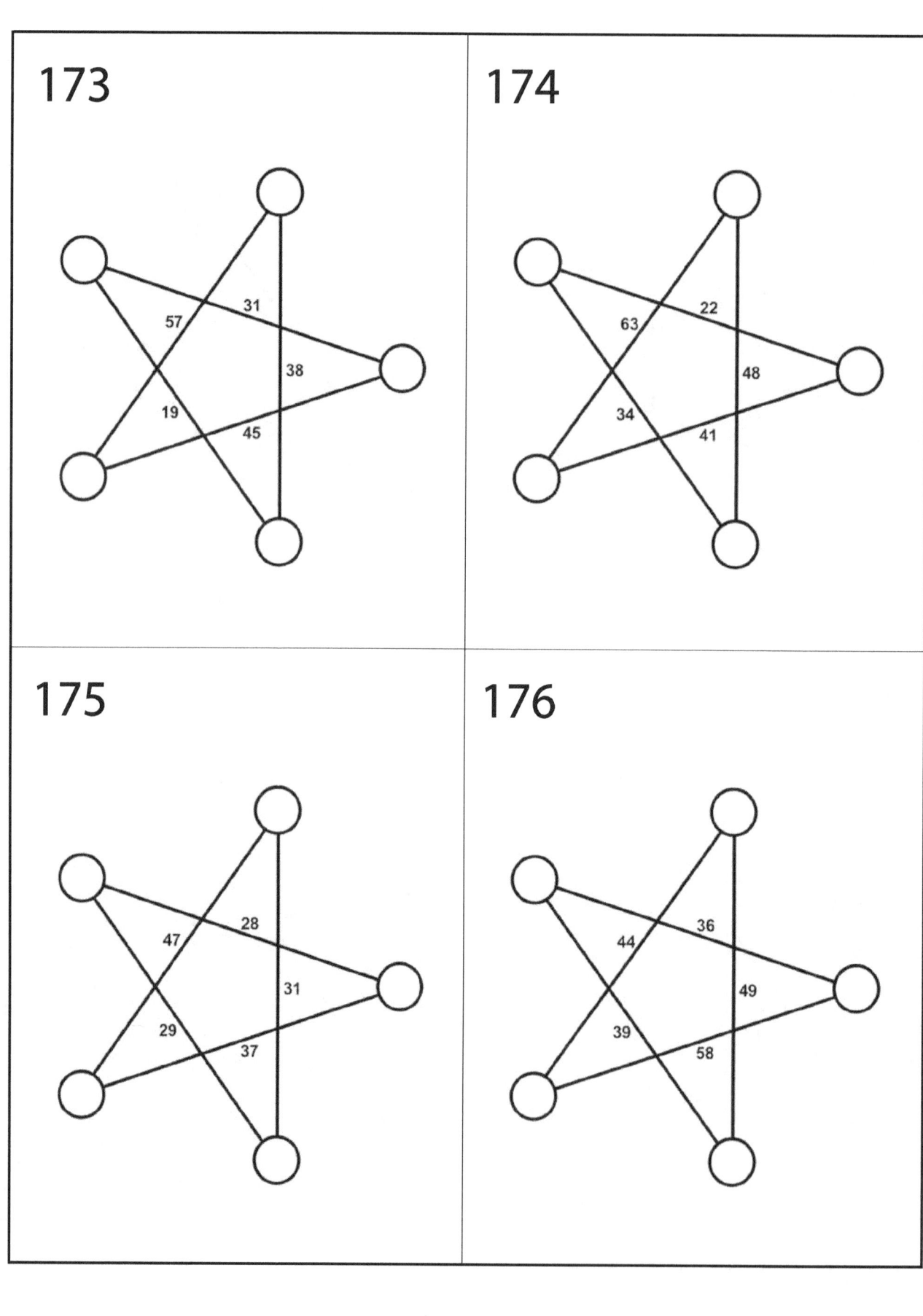

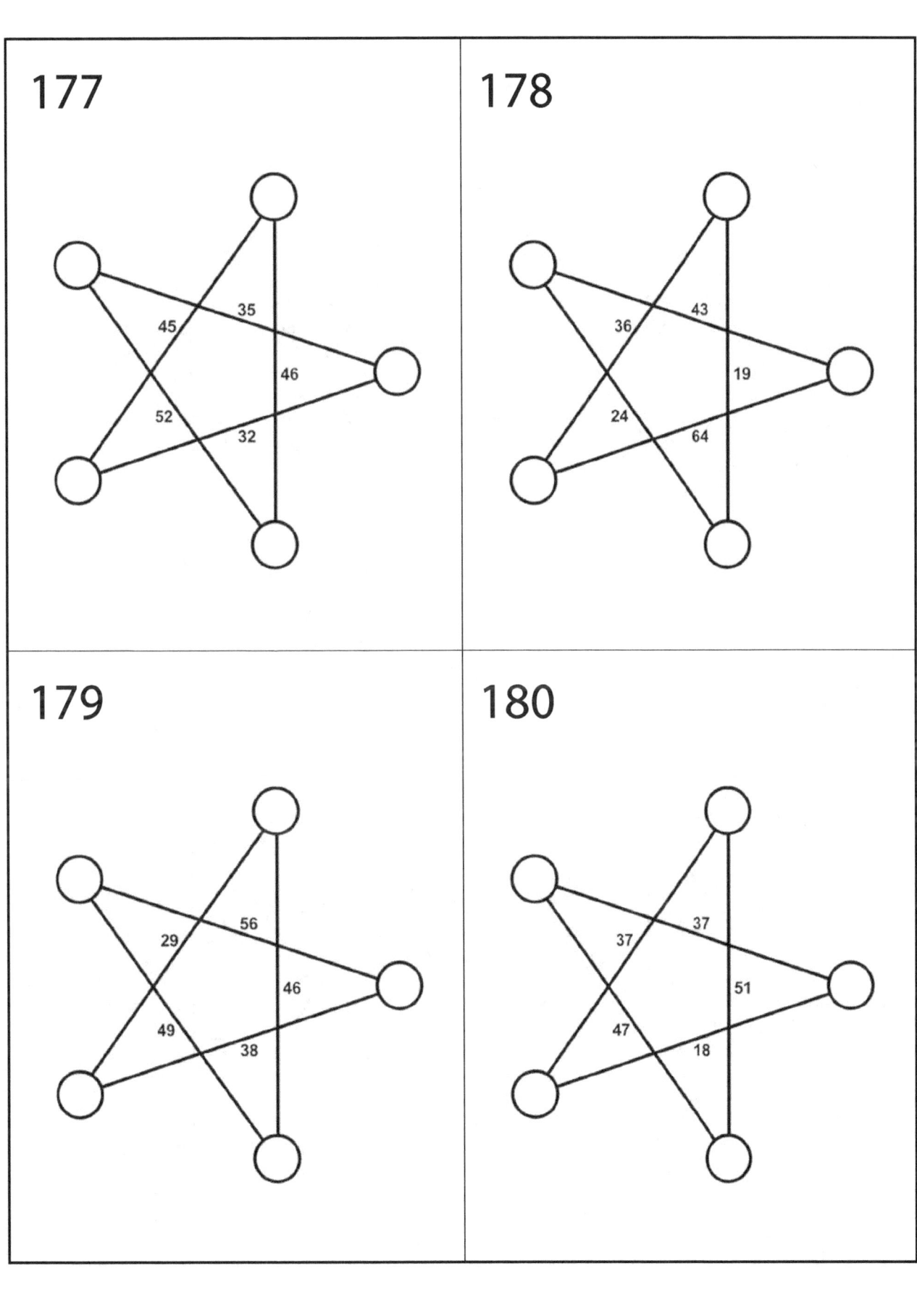

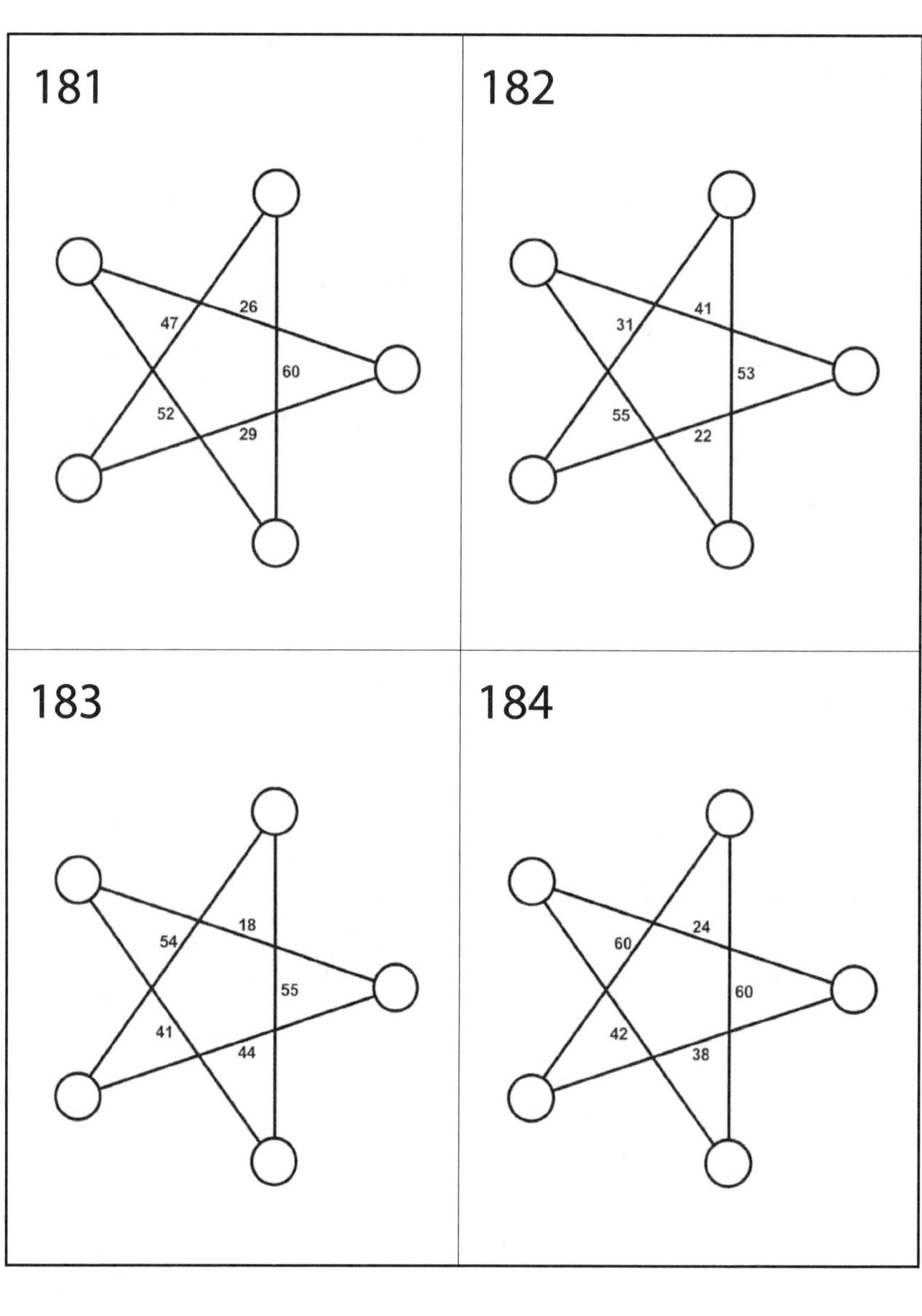

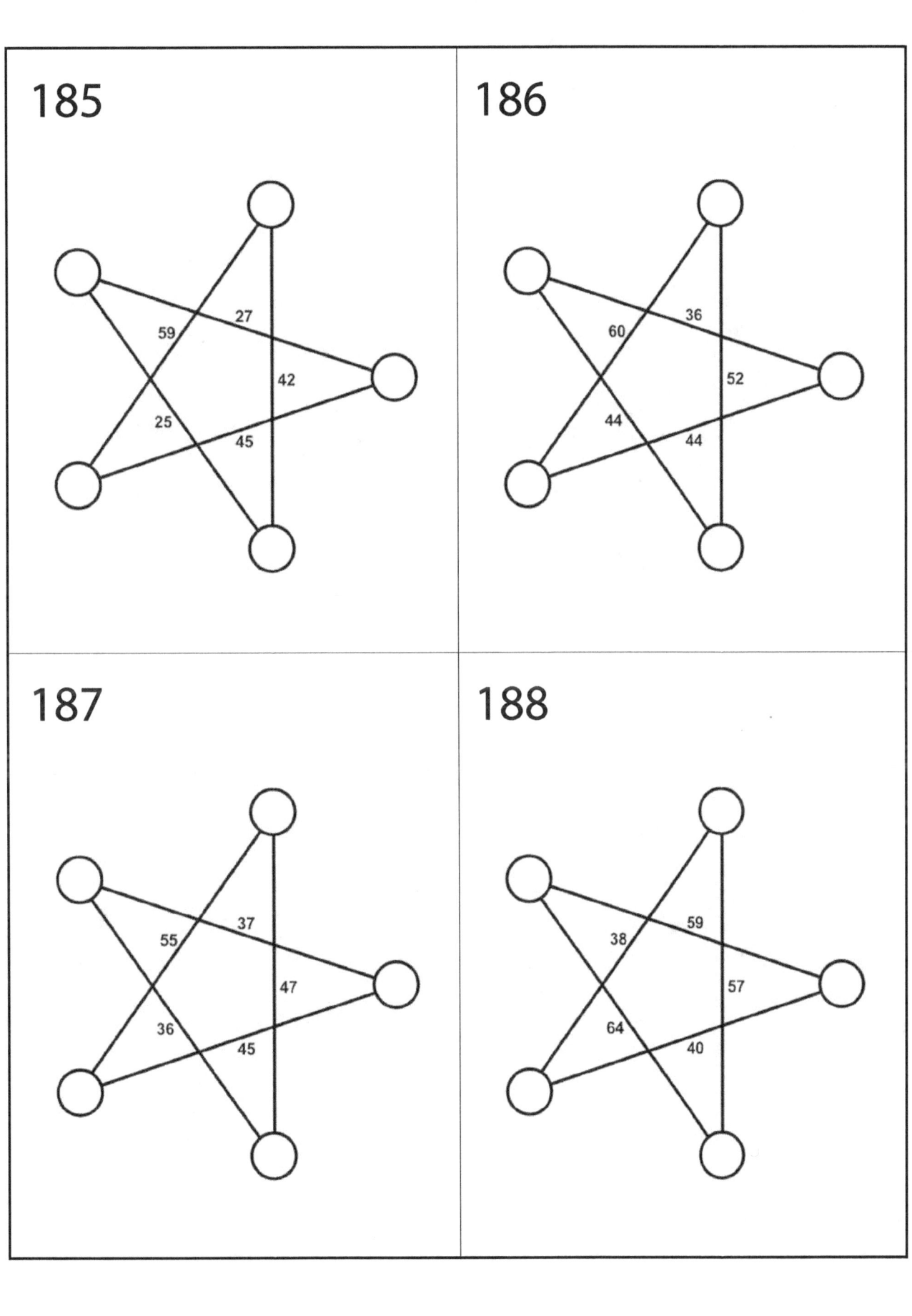

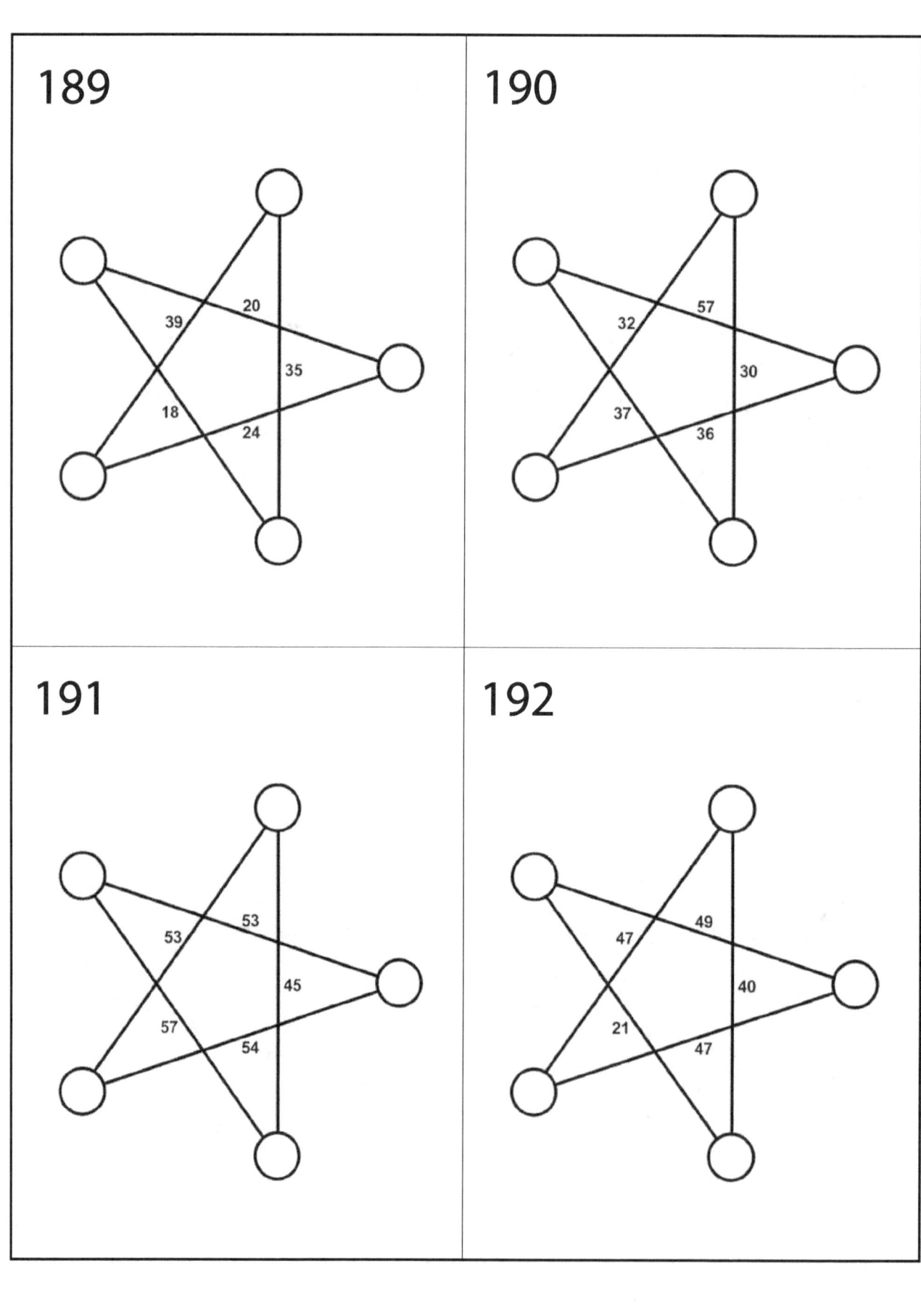

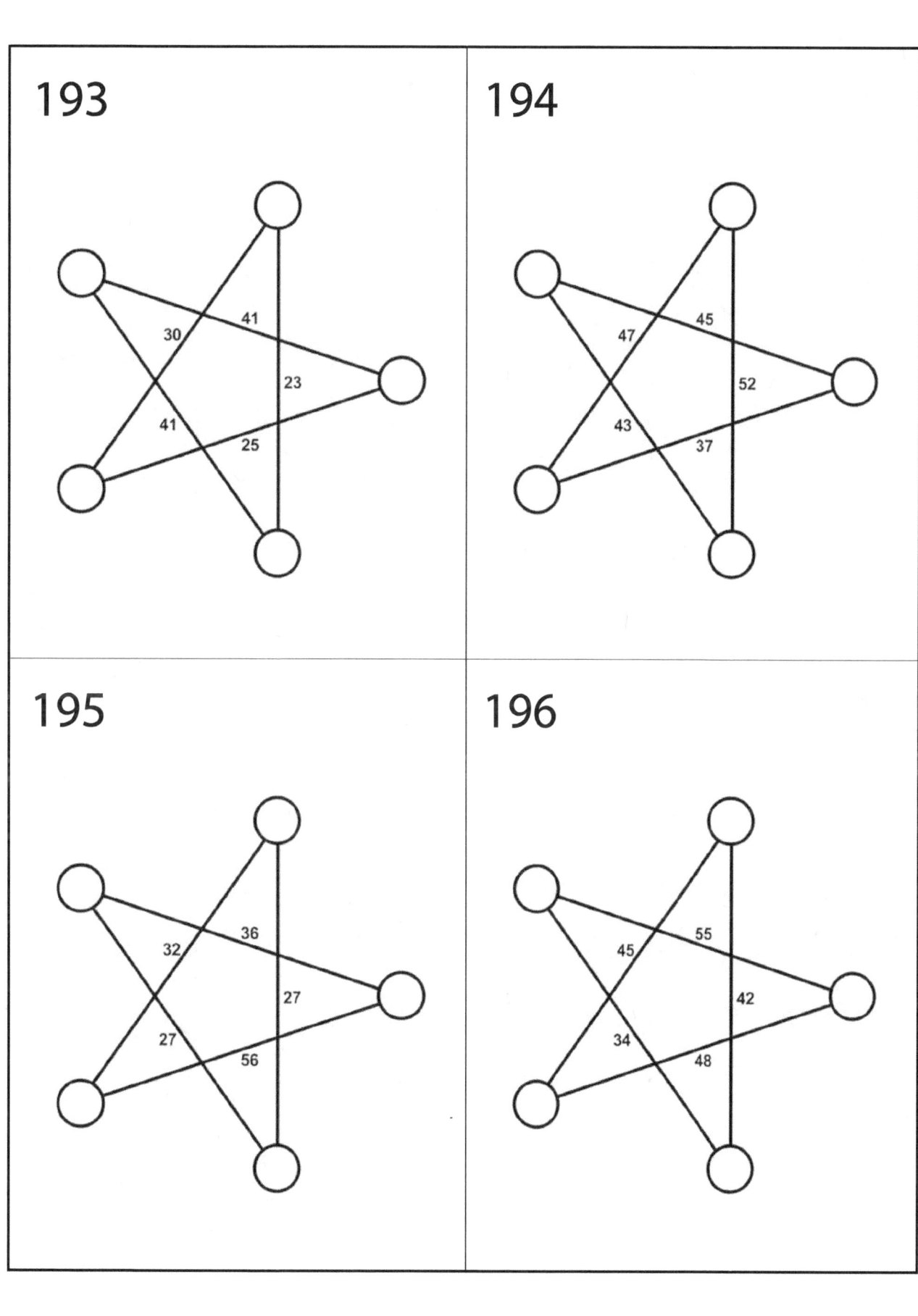

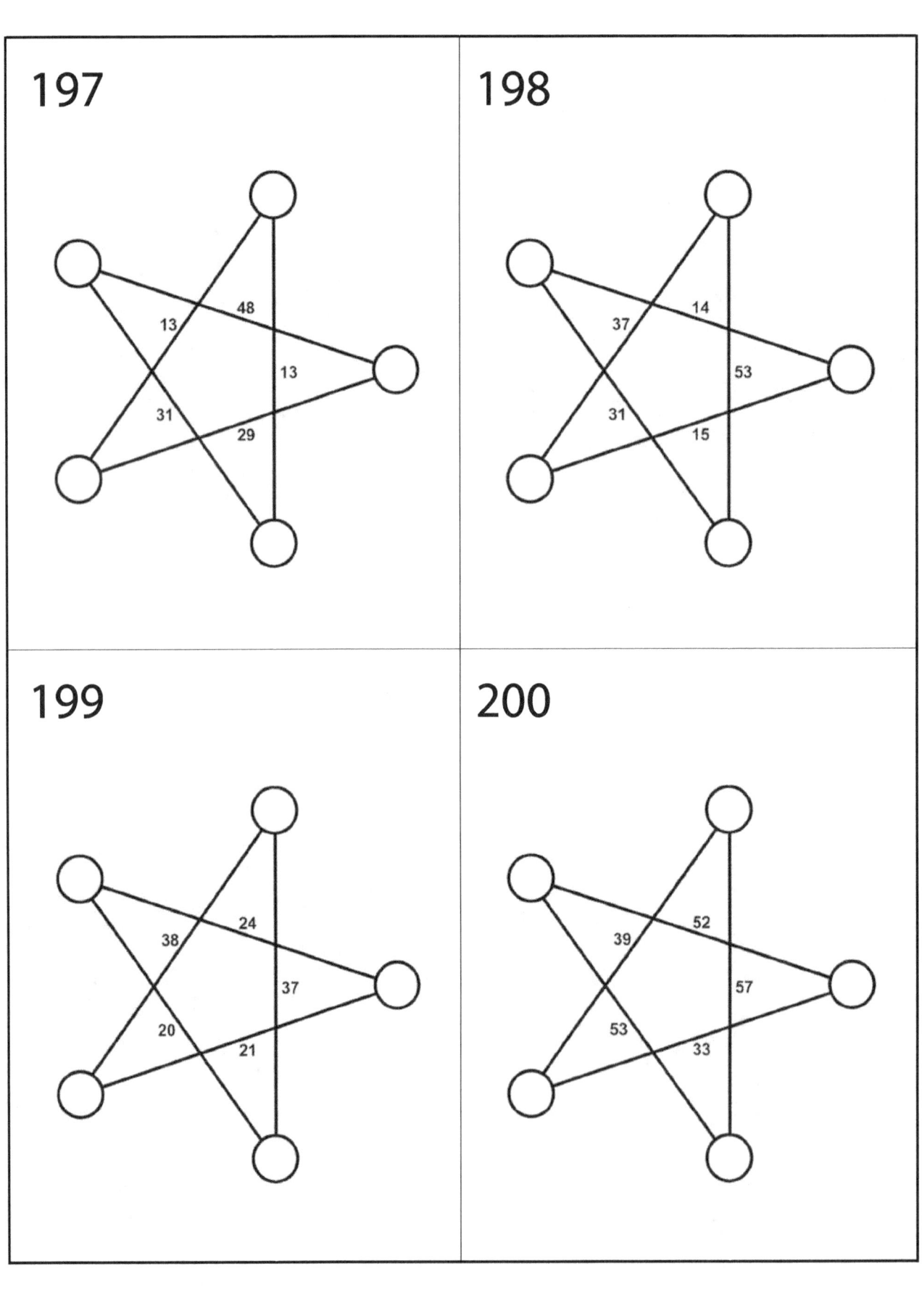

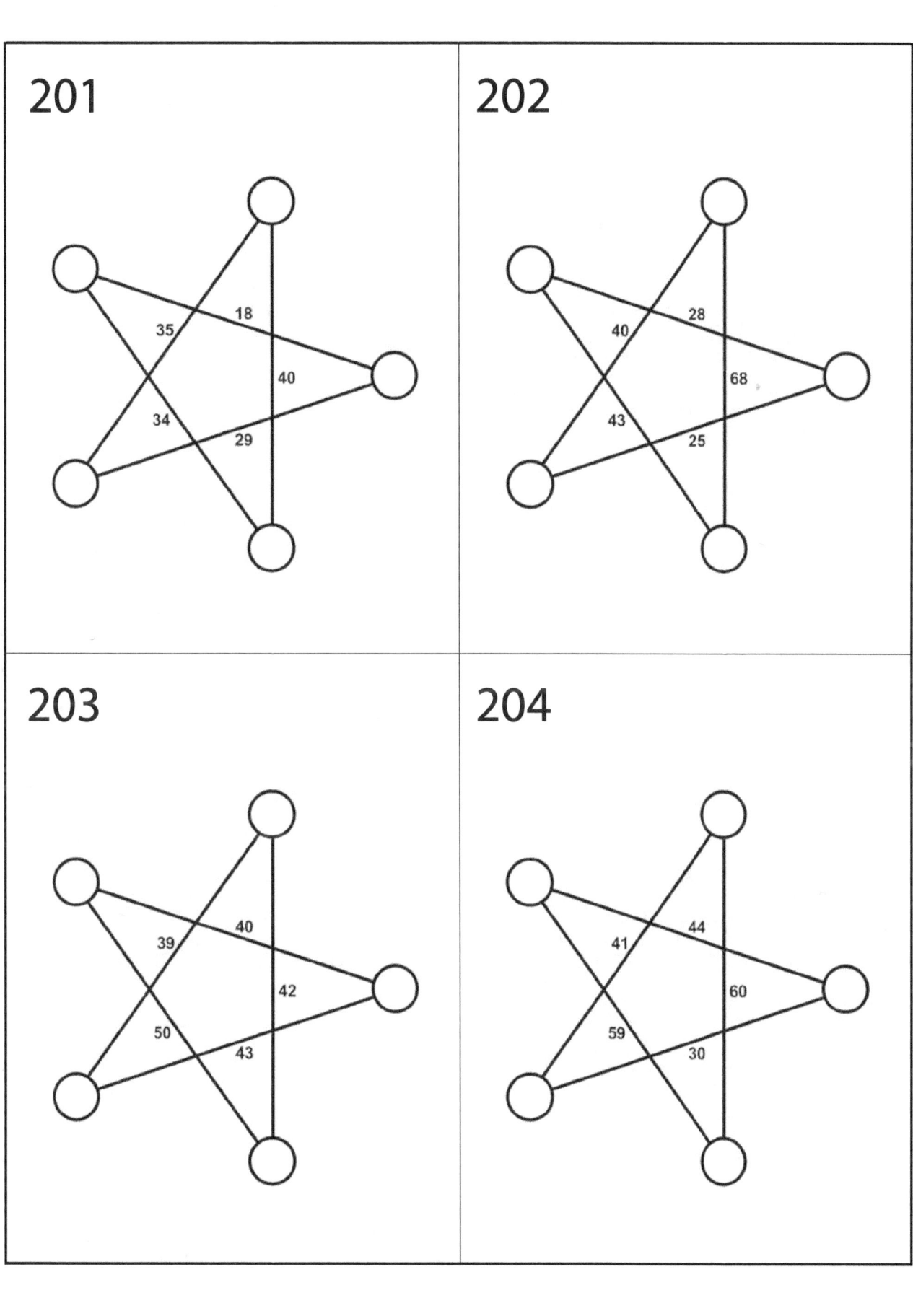

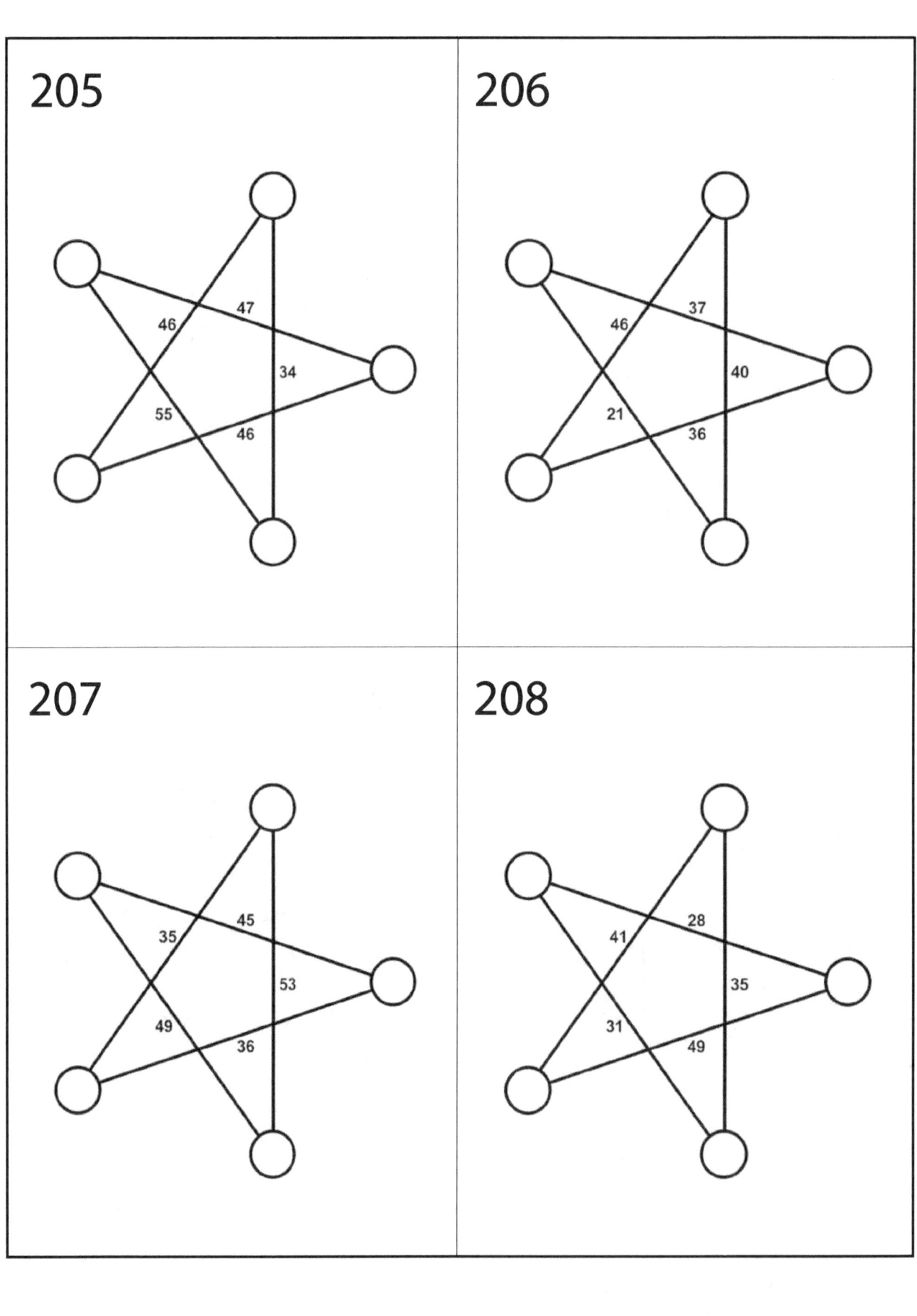

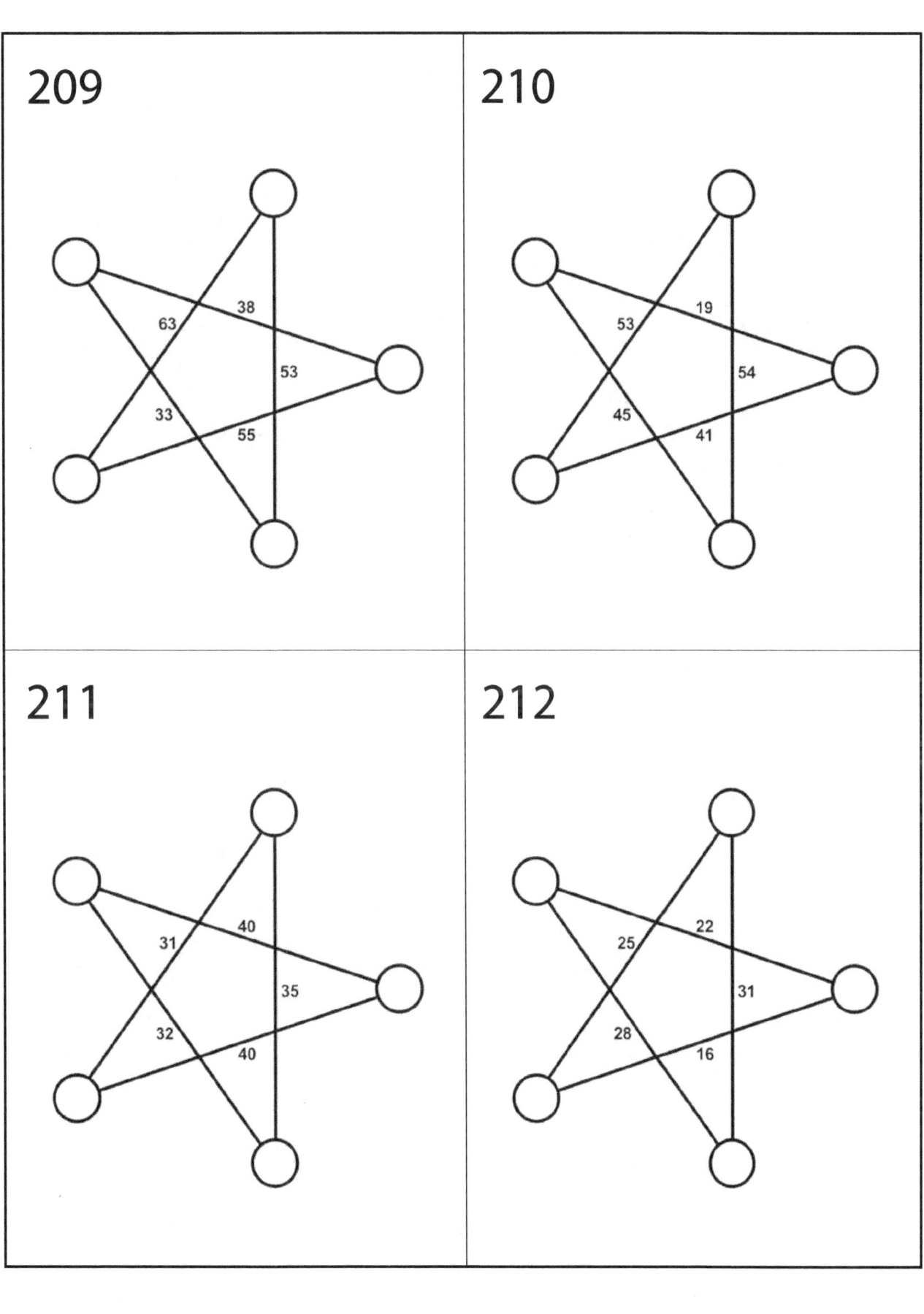

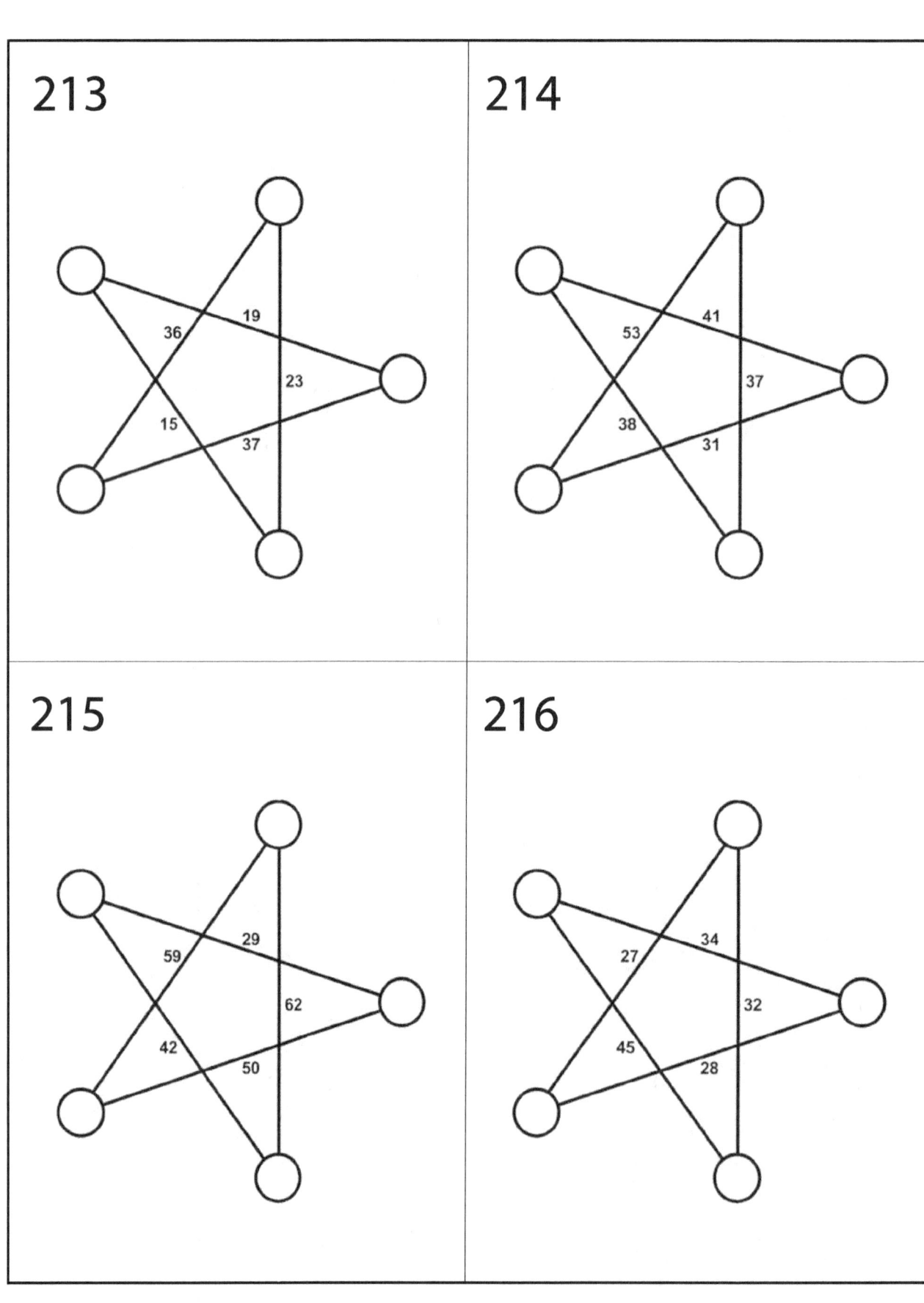

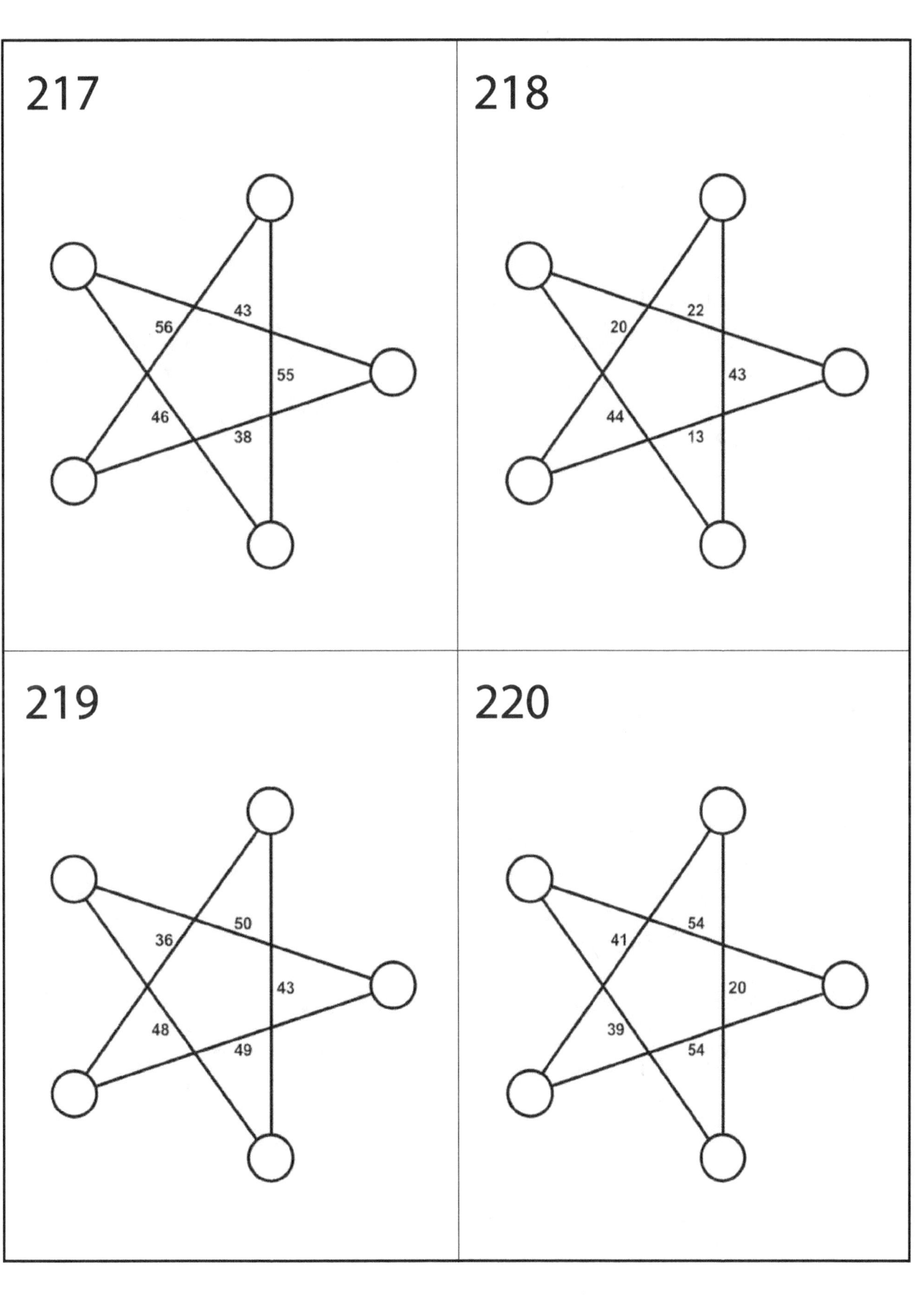

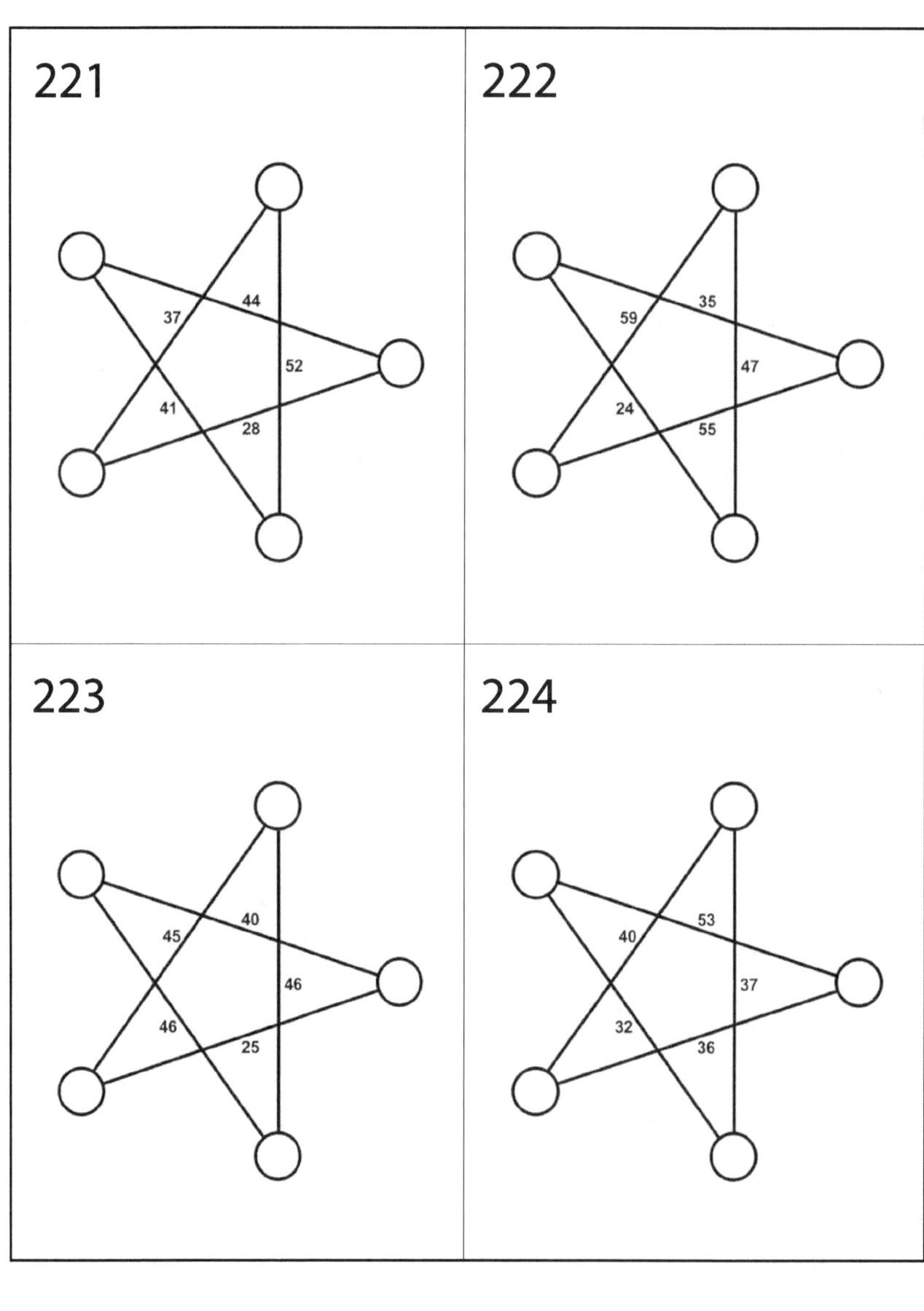

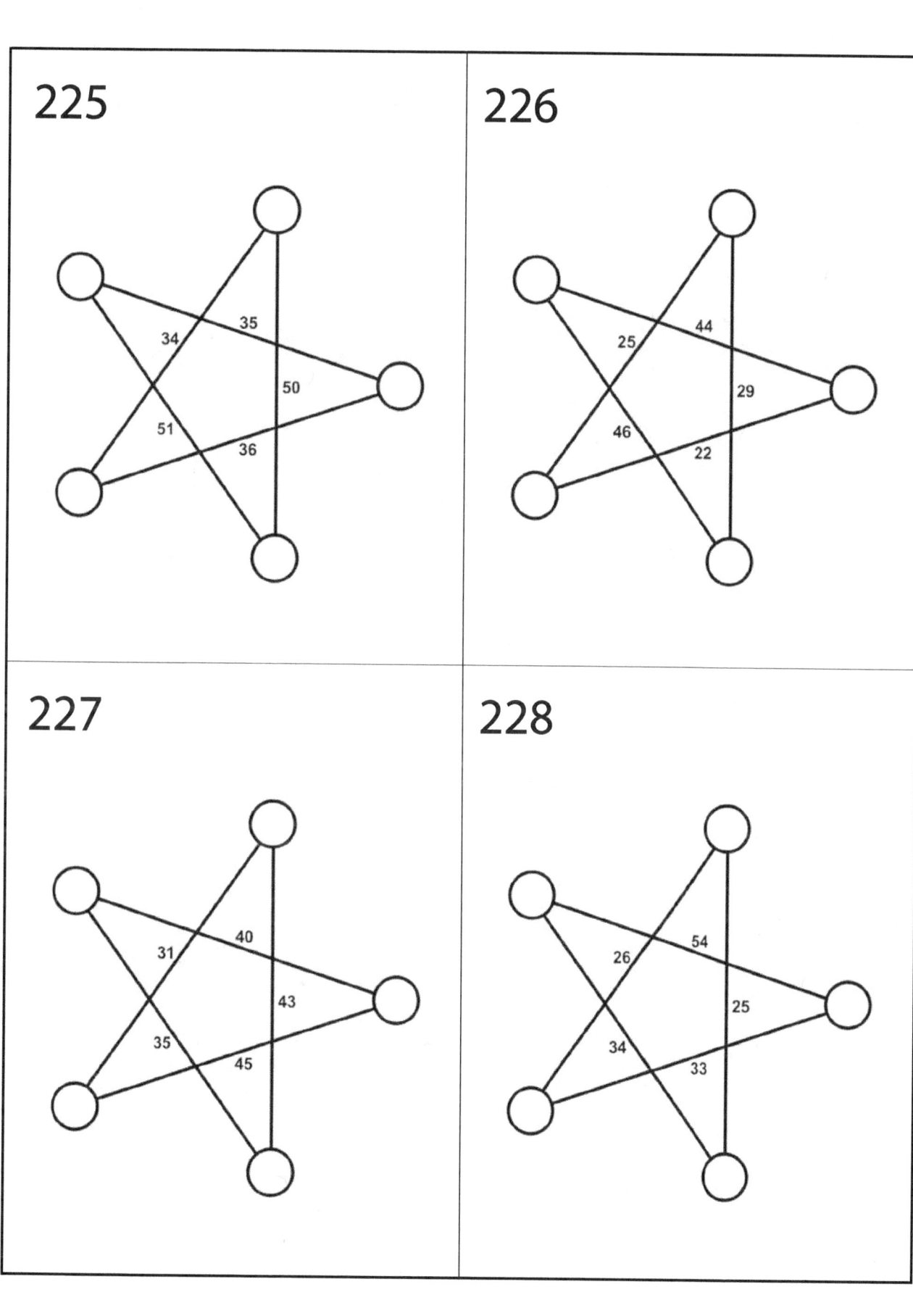

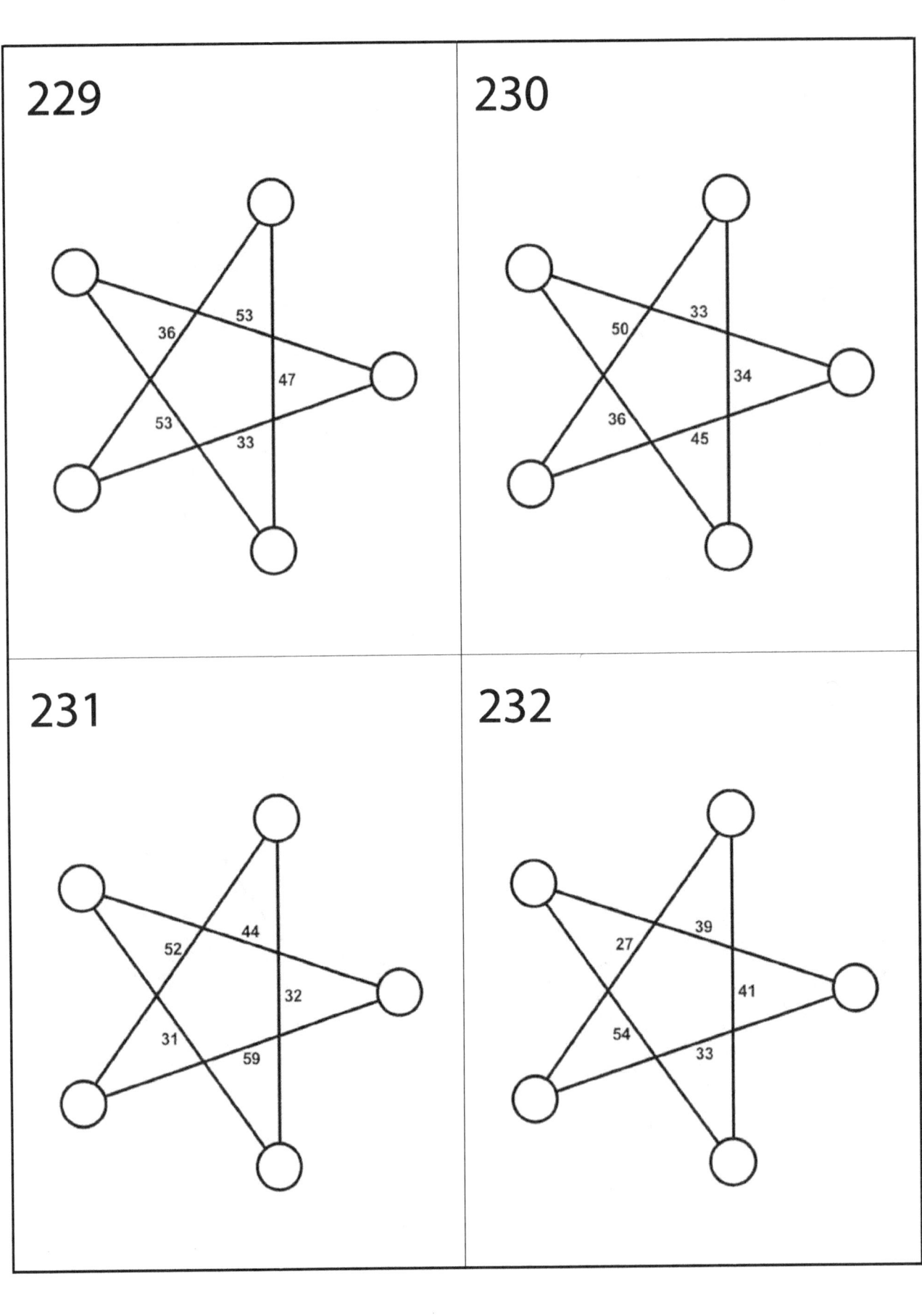

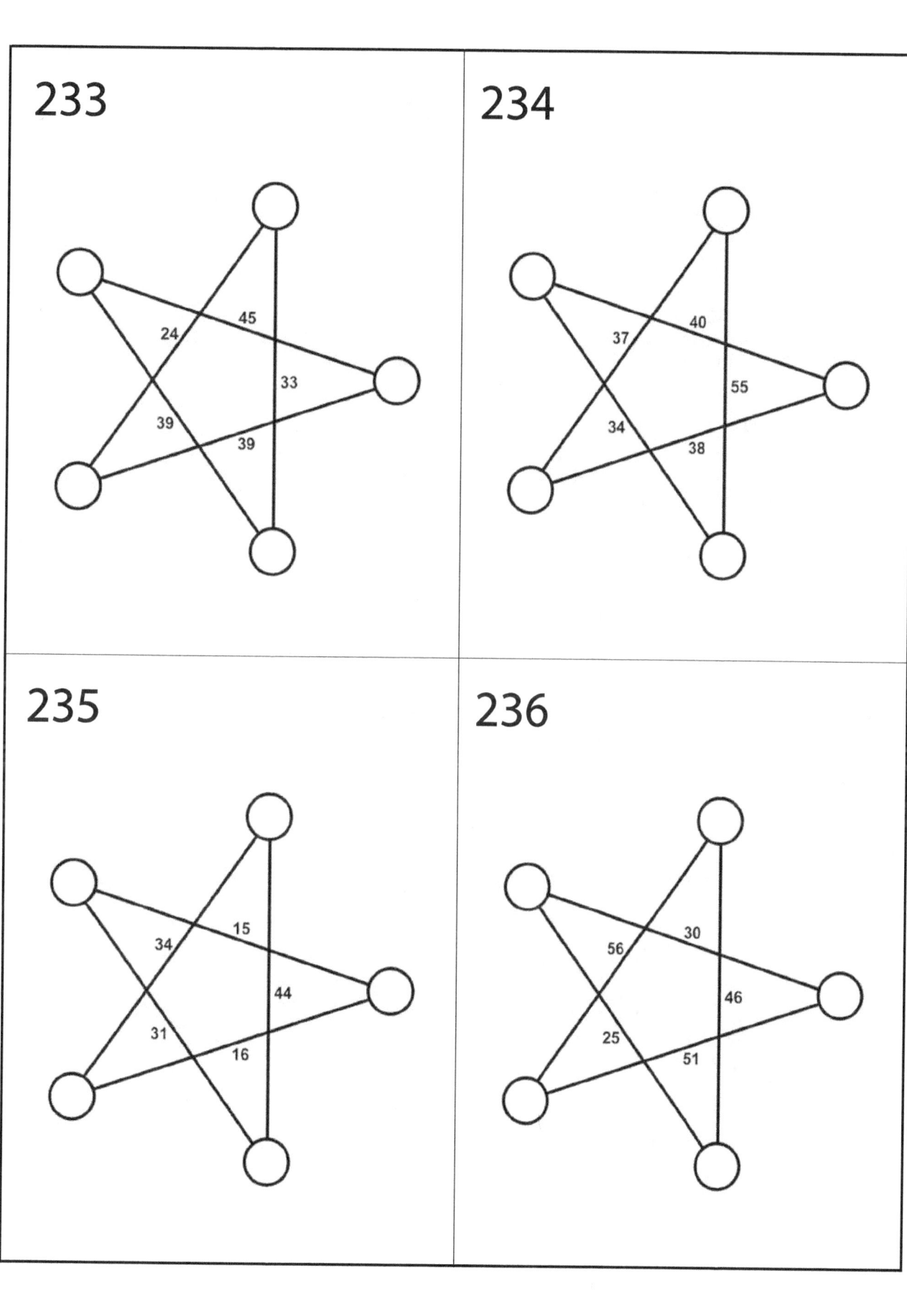

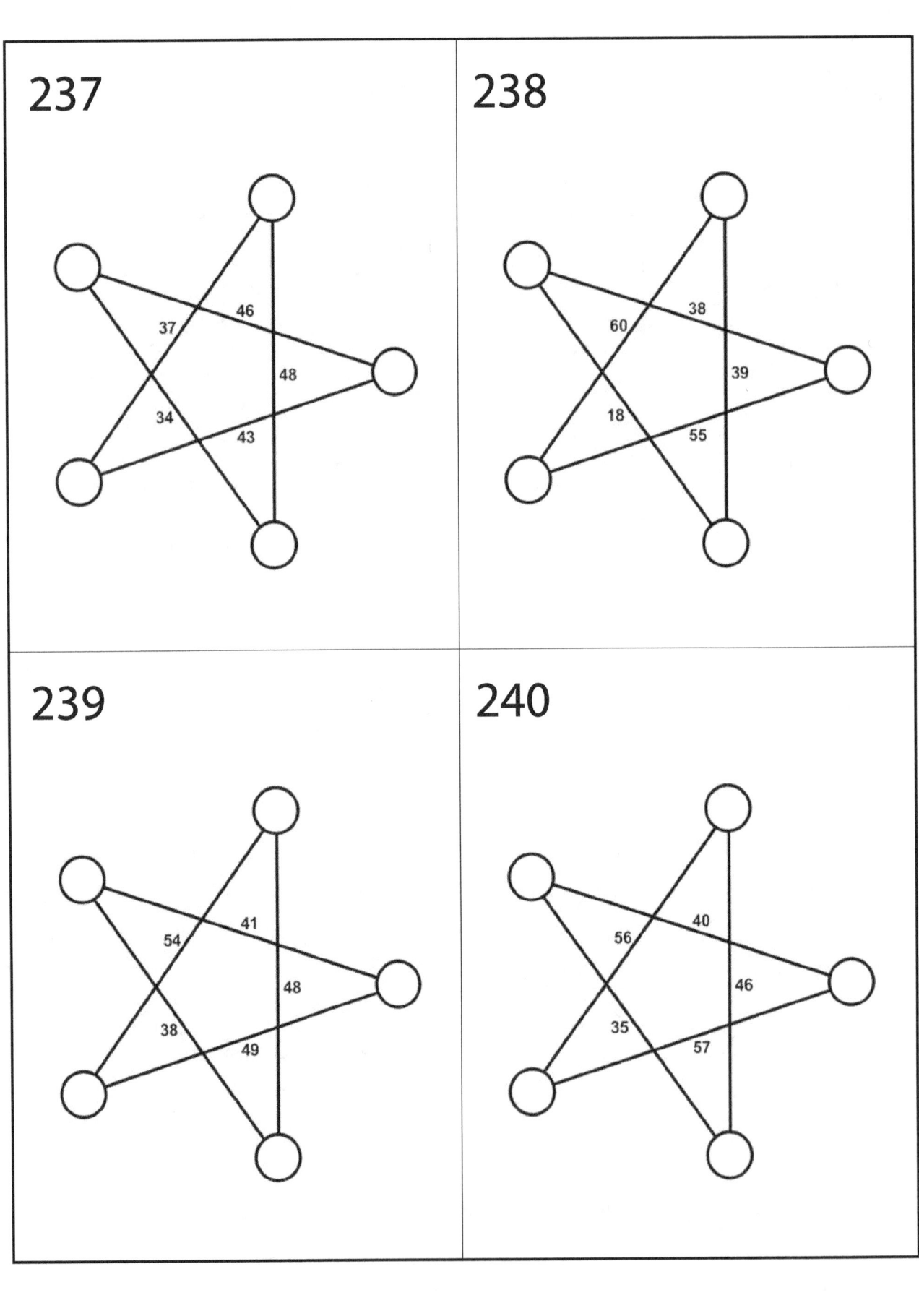

## 241

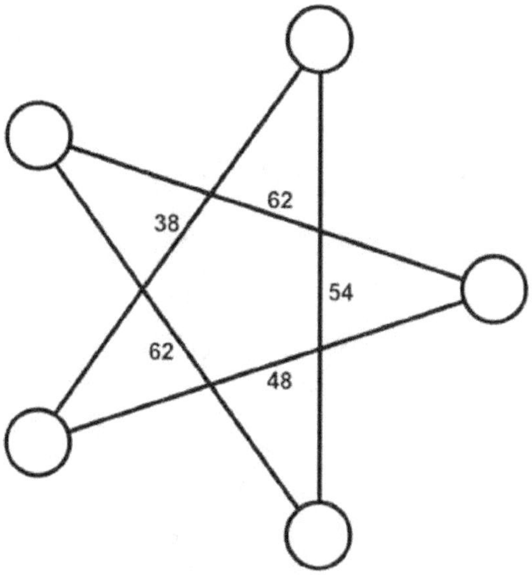

- 38
- 62
- 54
- 62
- 48

## 242

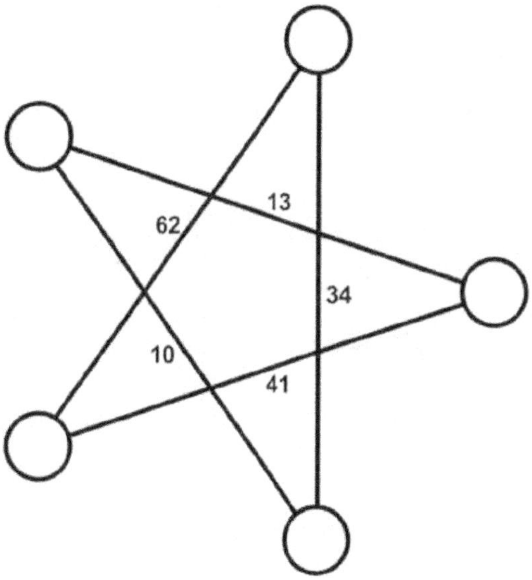

- 62
- 13
- 34
- 10
- 41

## 243

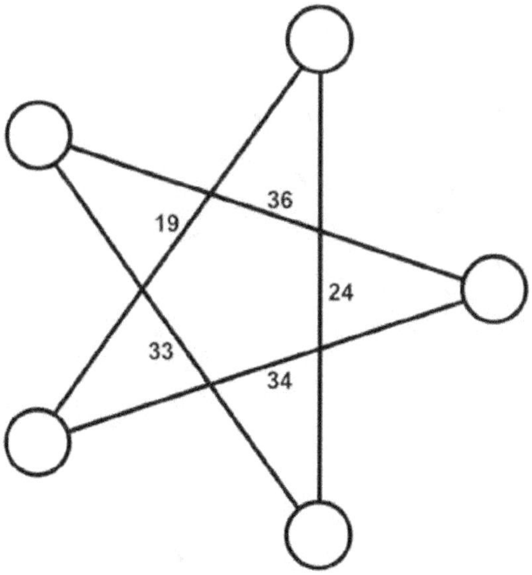

- 19
- 36
- 24
- 33
- 34

## 244

- 47
- 46
- 29
- 46
- 48

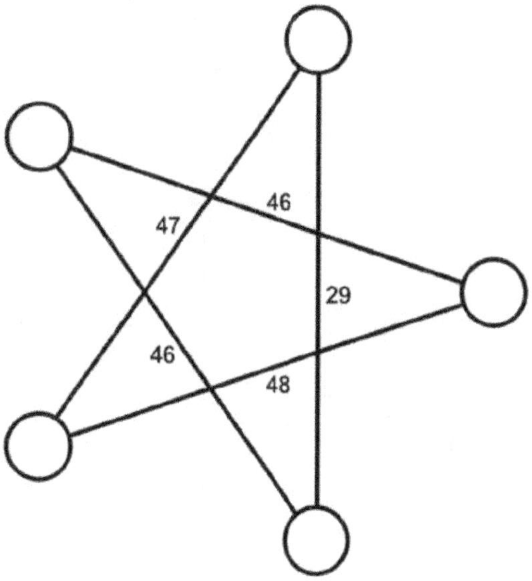

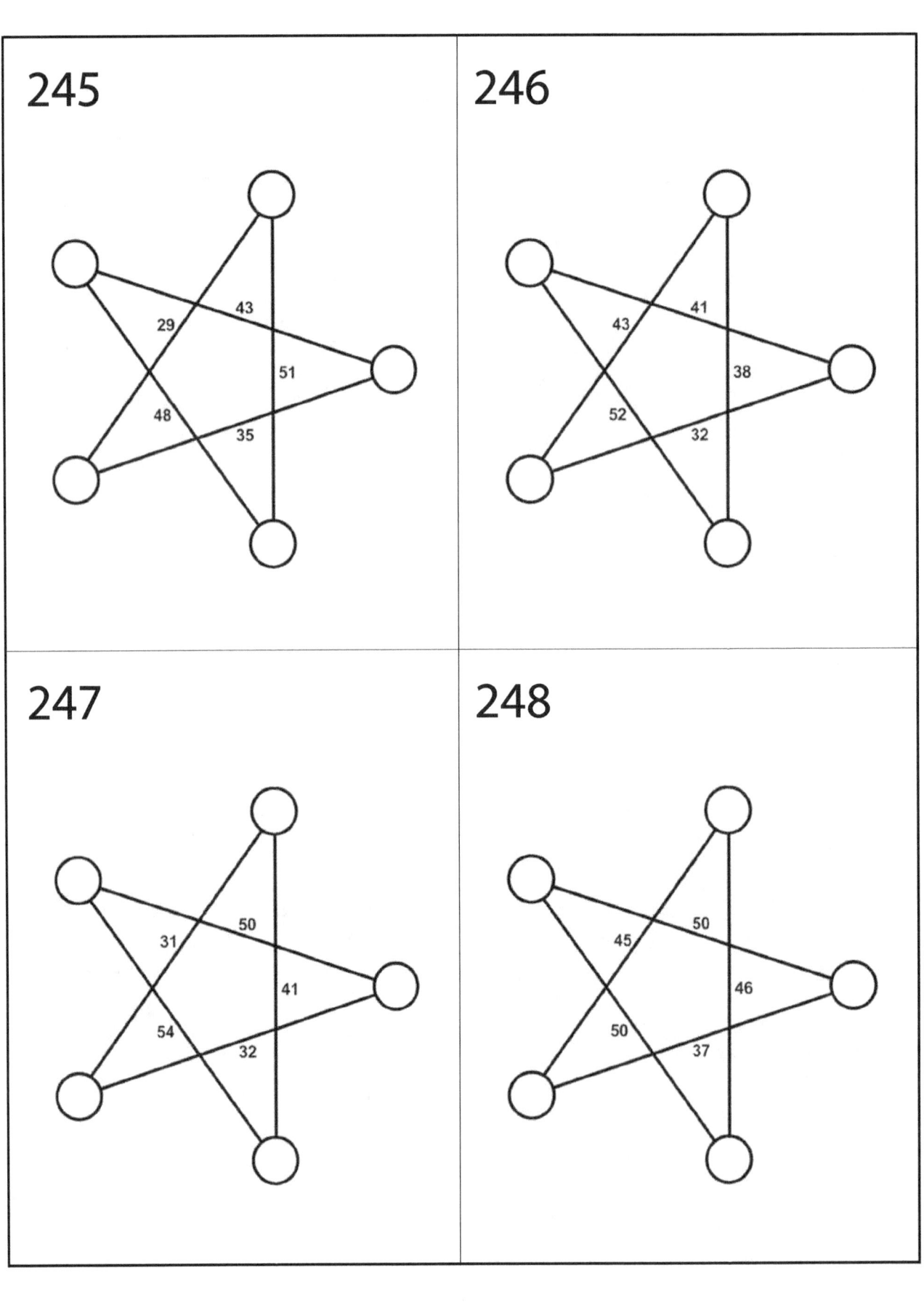

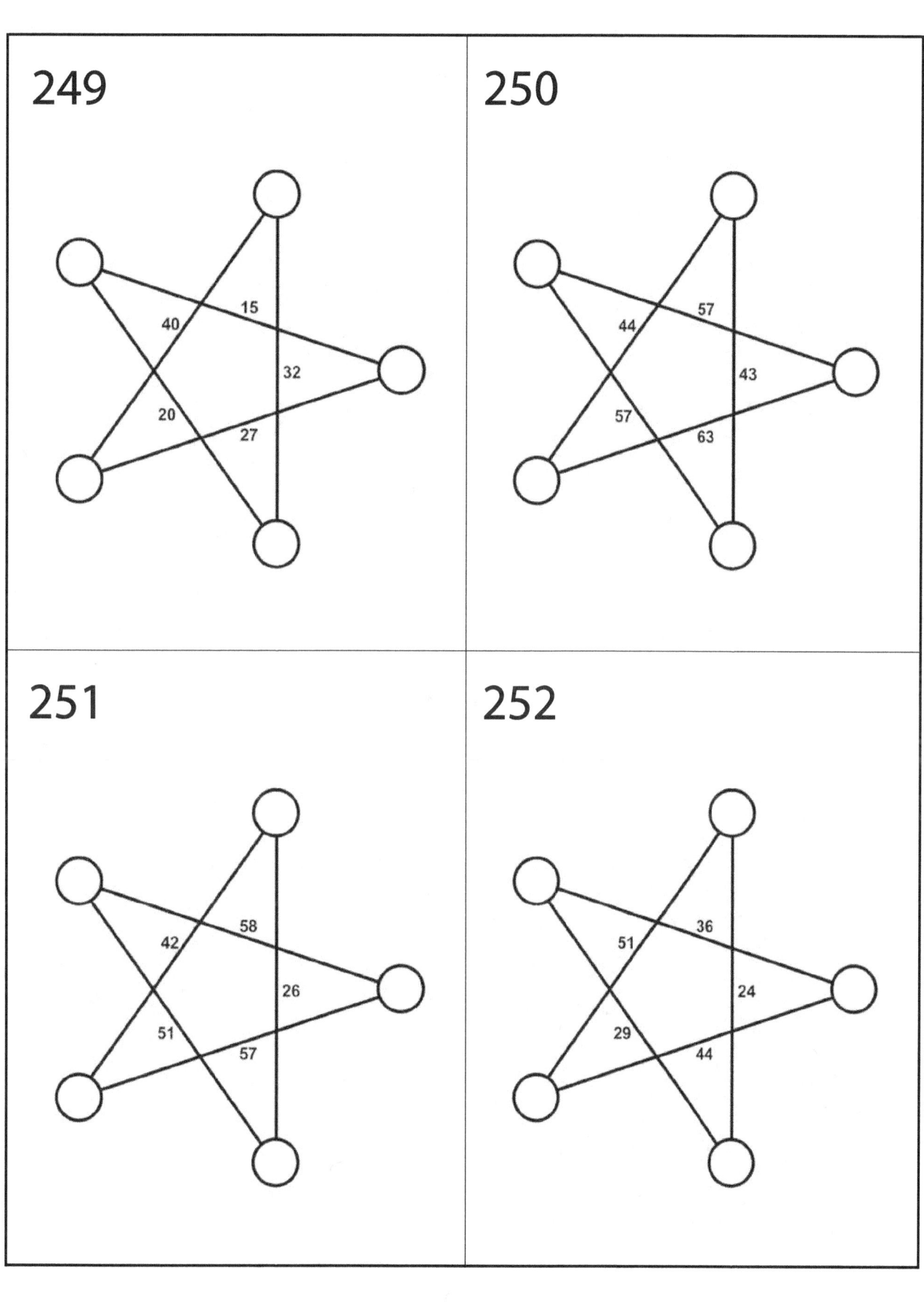

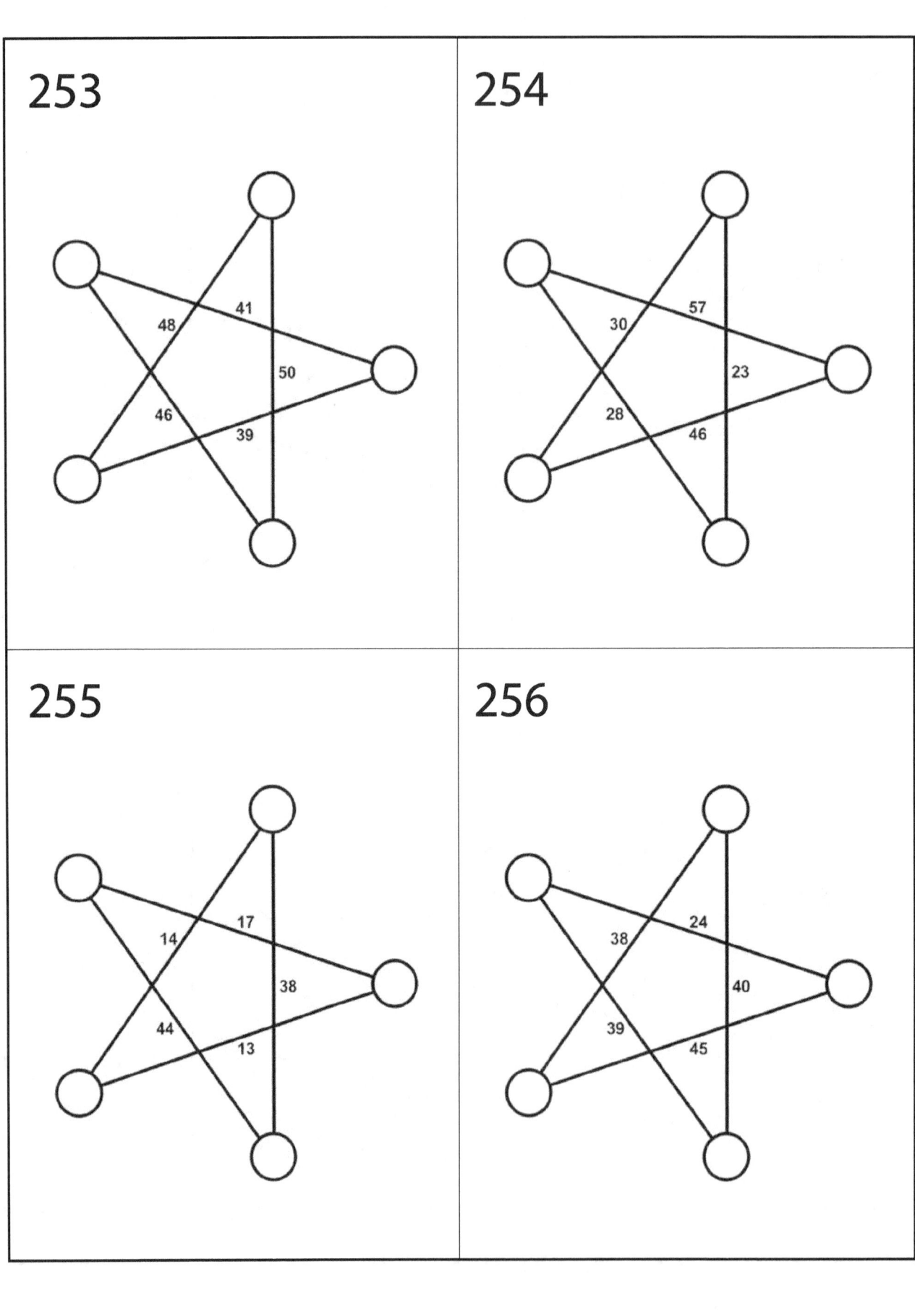

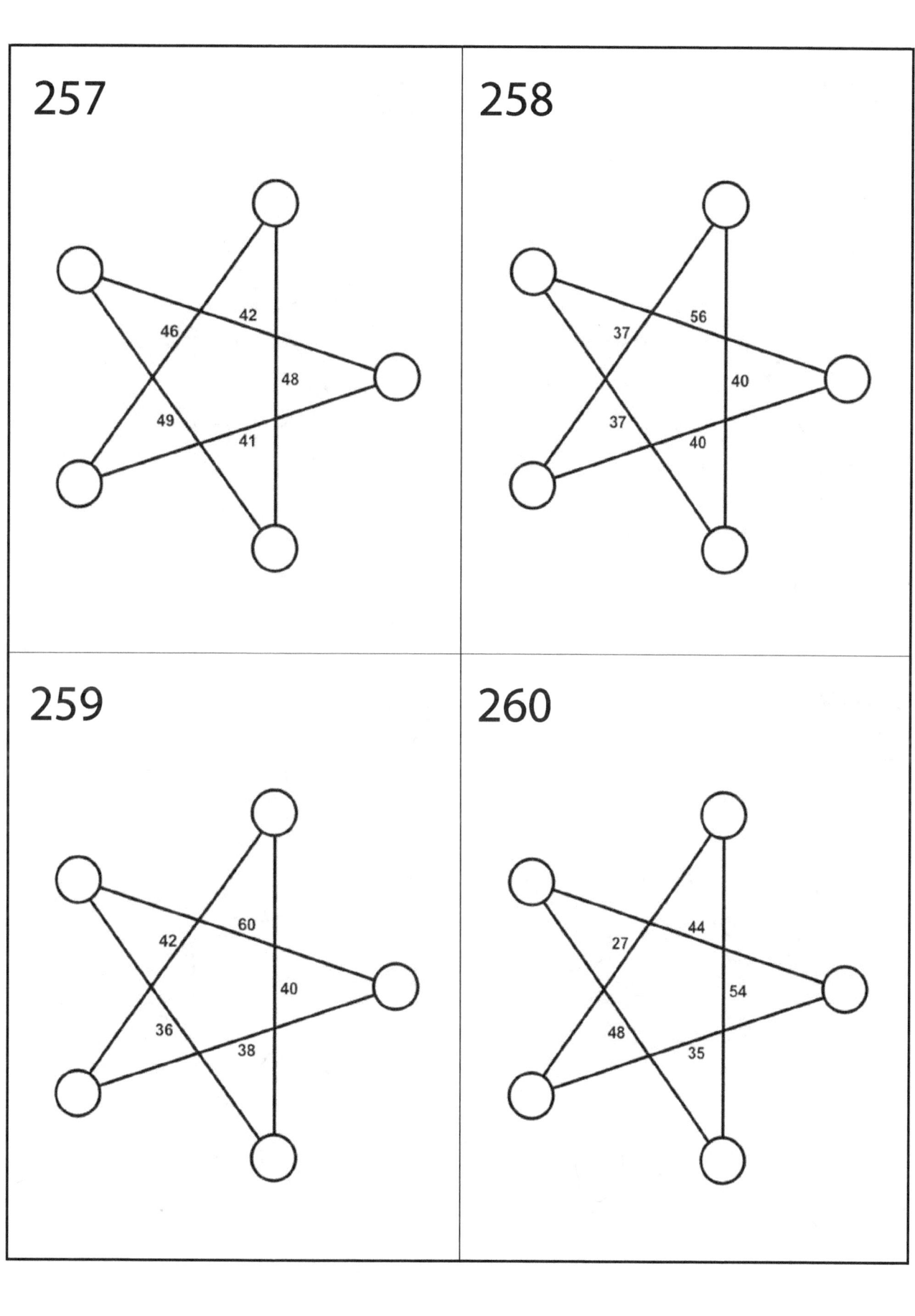

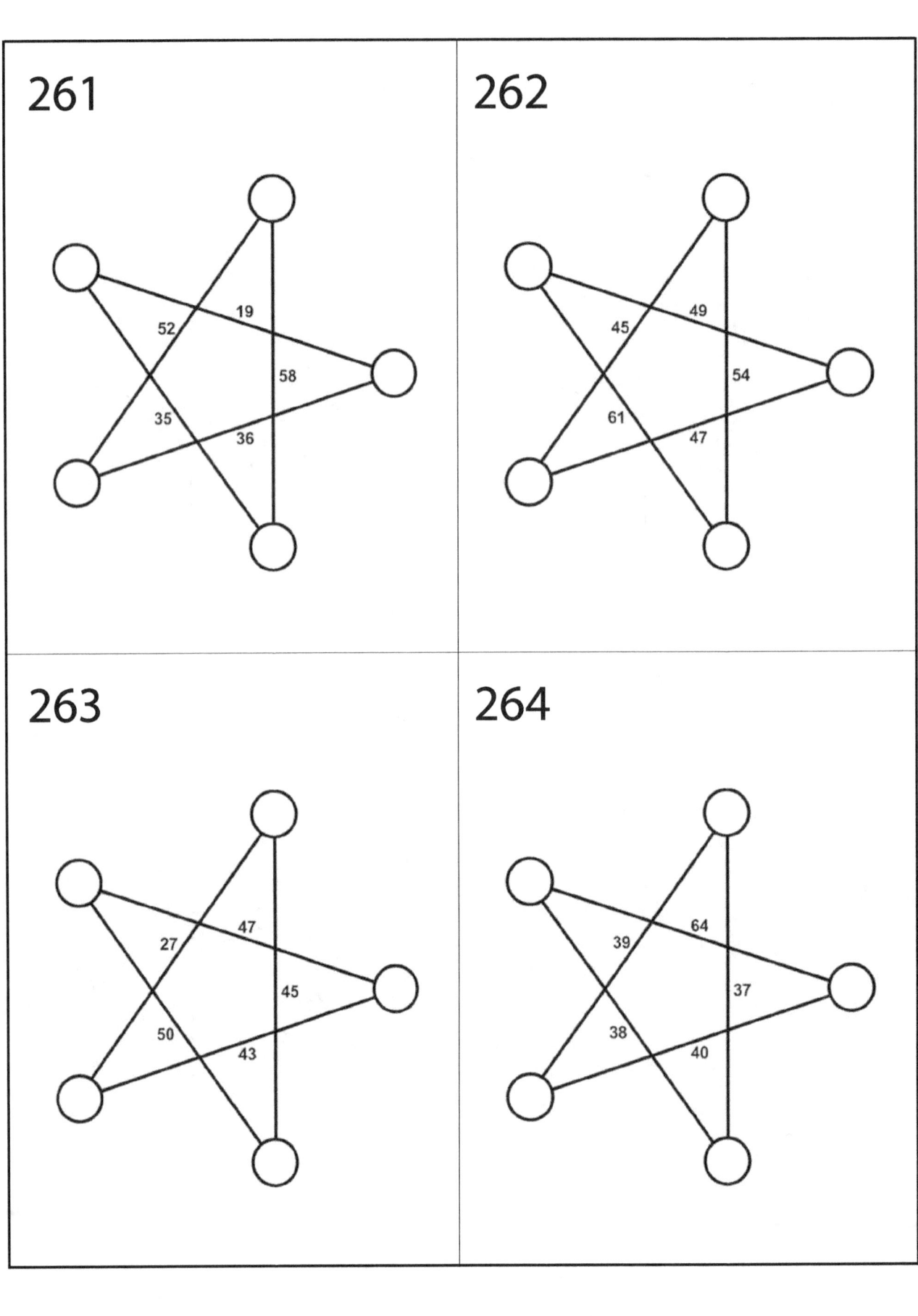

## 265

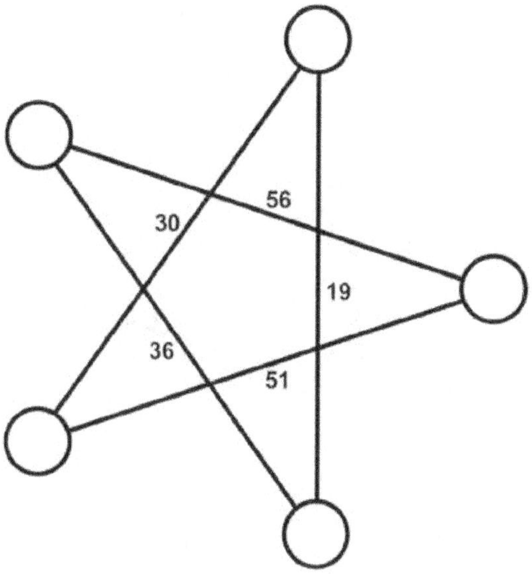

- 30
- 56
- 19
- 36
- 51

## 266

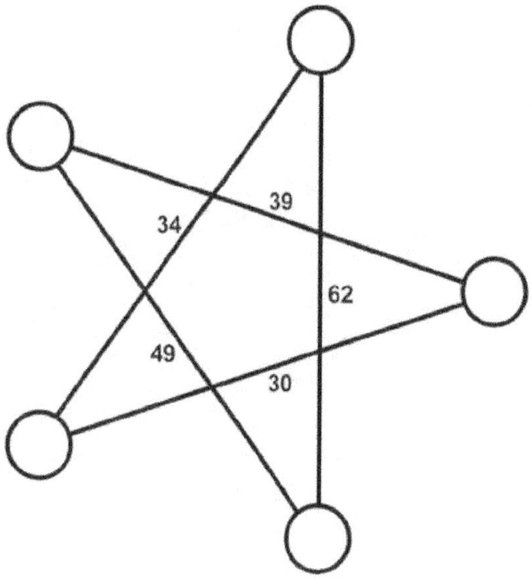

- 34
- 39
- 62
- 49
- 30

## 267

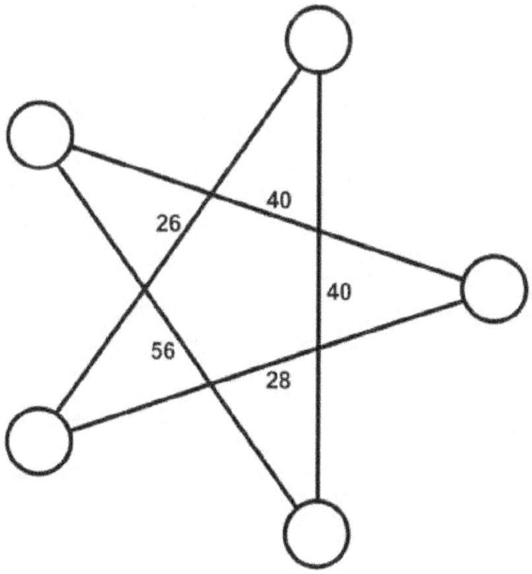

- 26
- 40
- 40
- 56
- 28

## 268

- 40
- 26
- 26
- 14
- 42

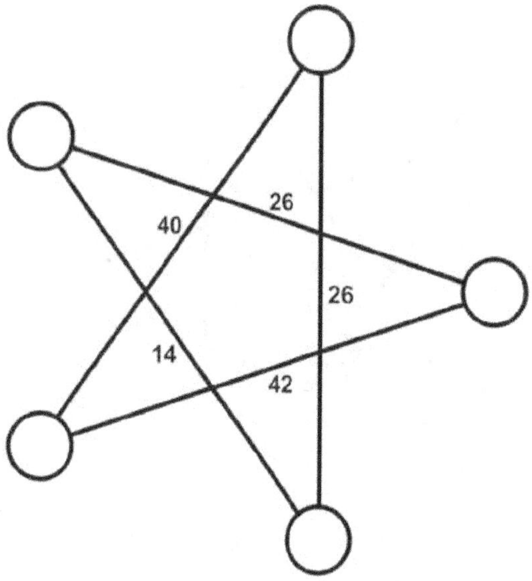

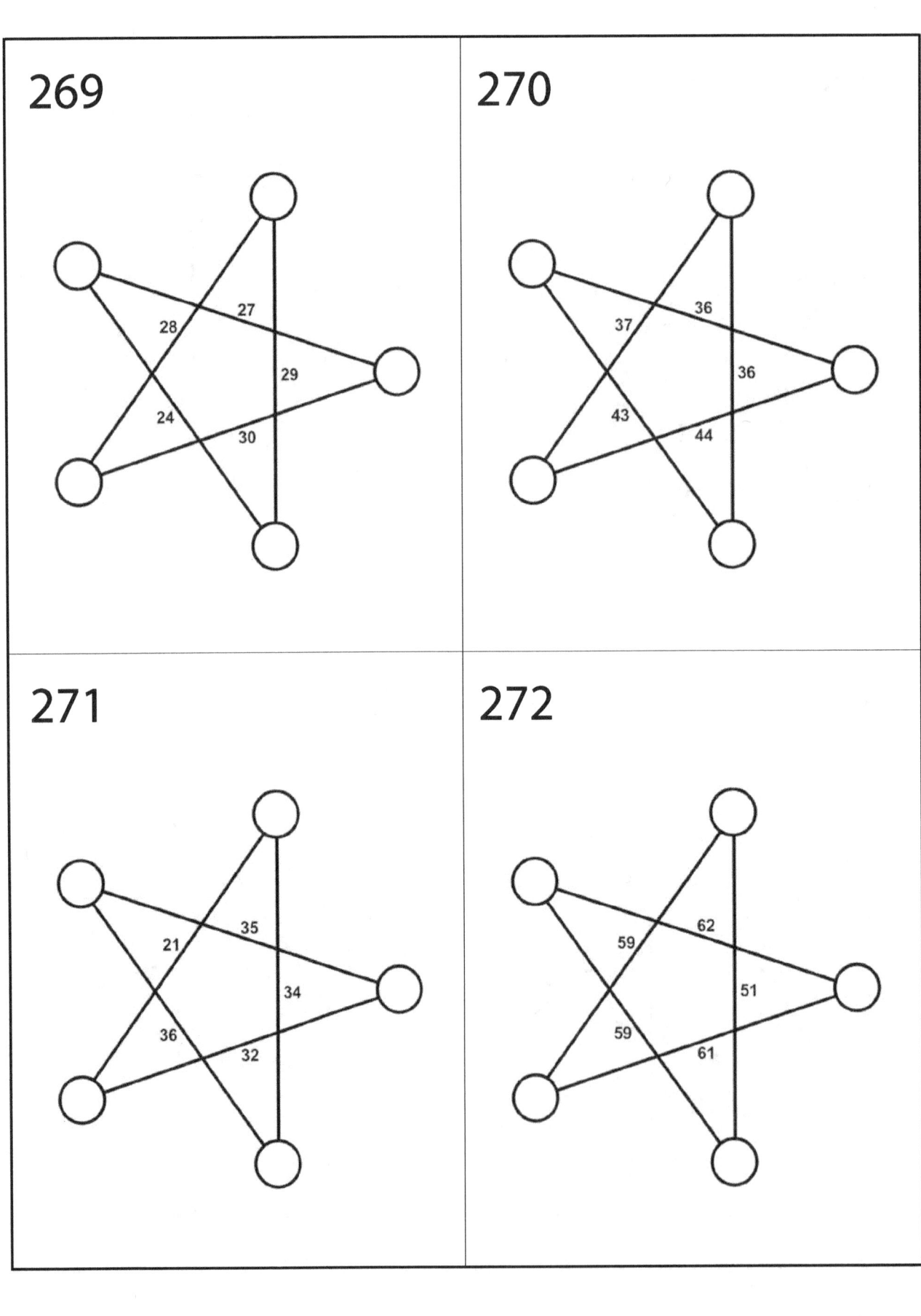

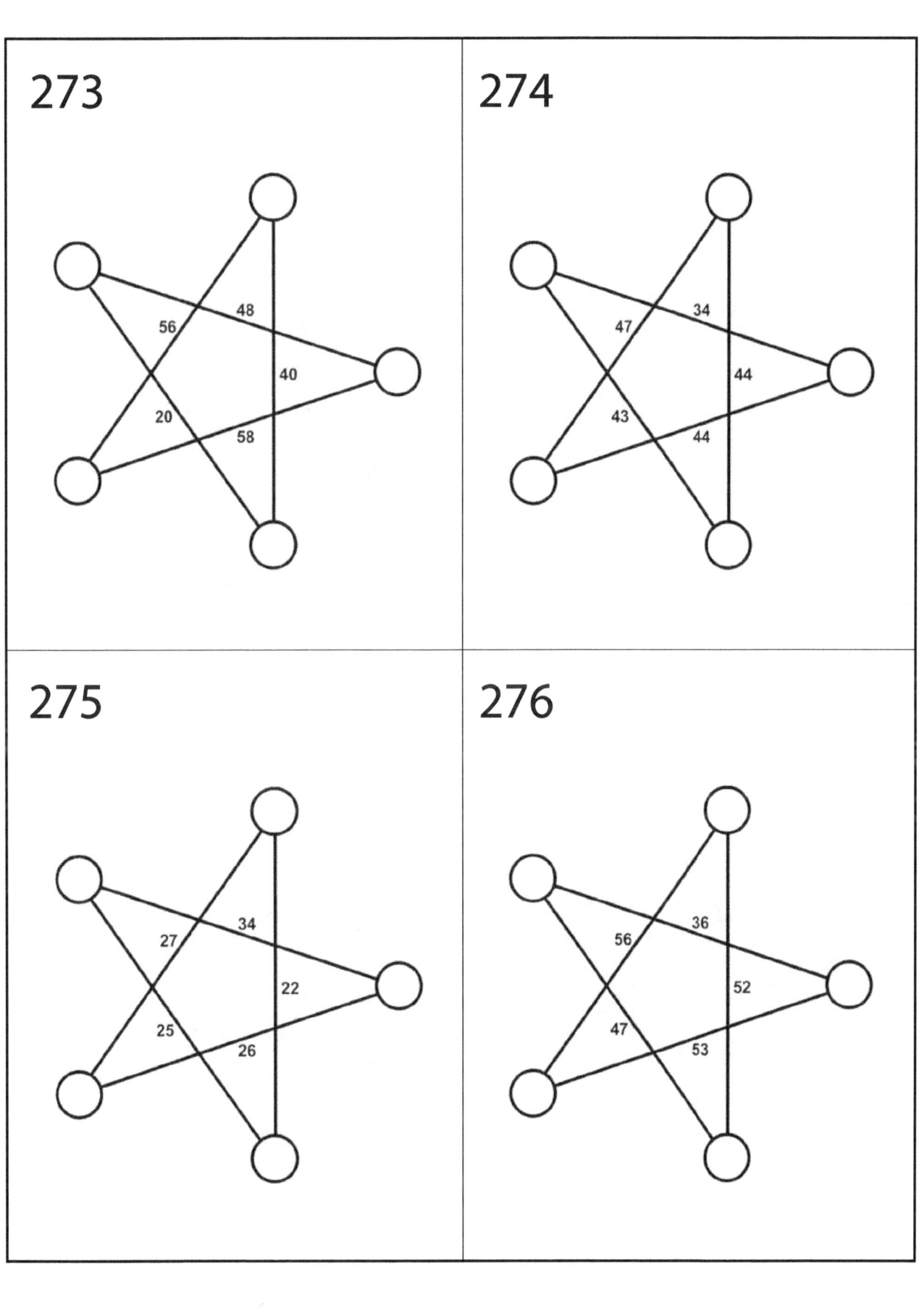

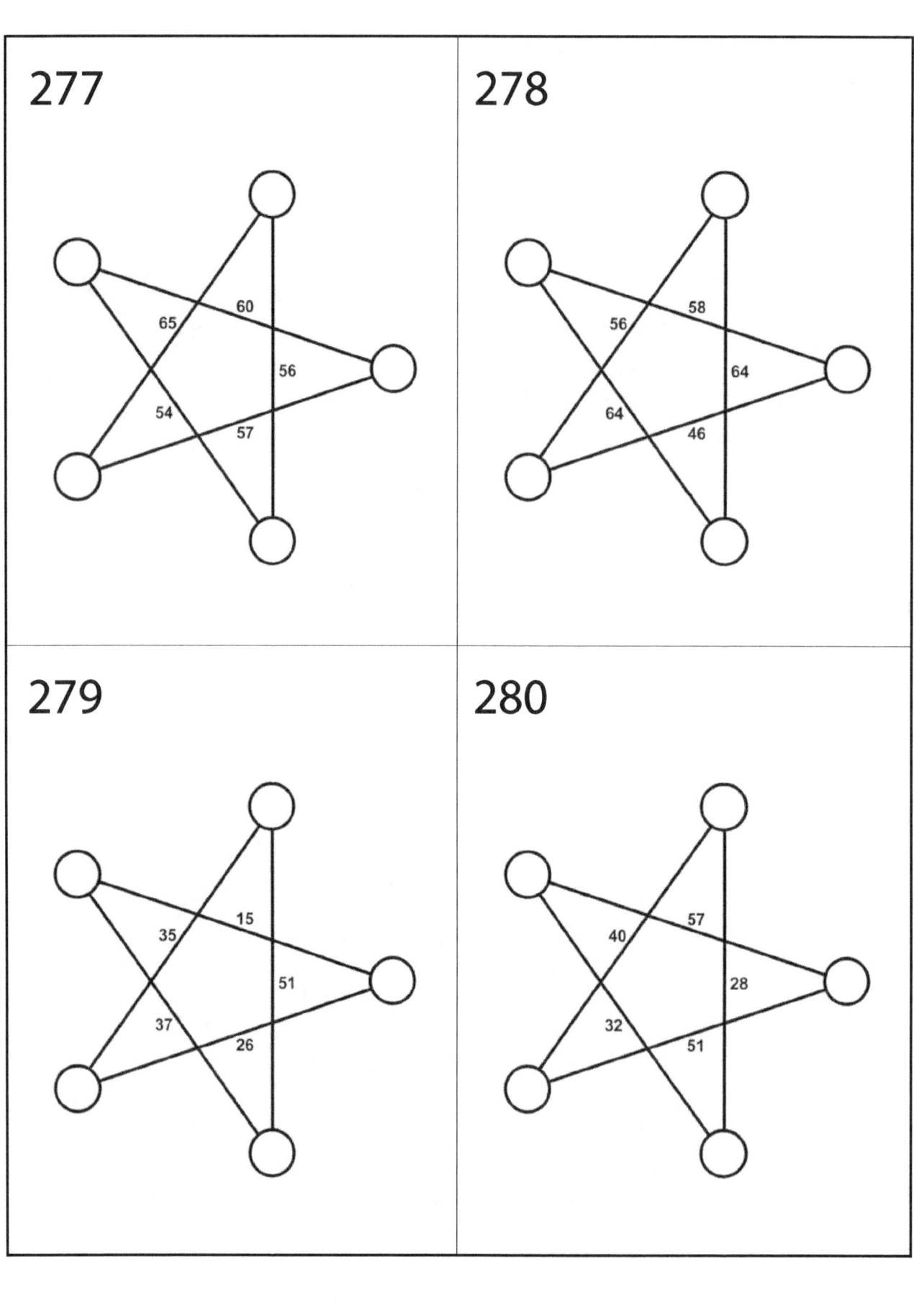

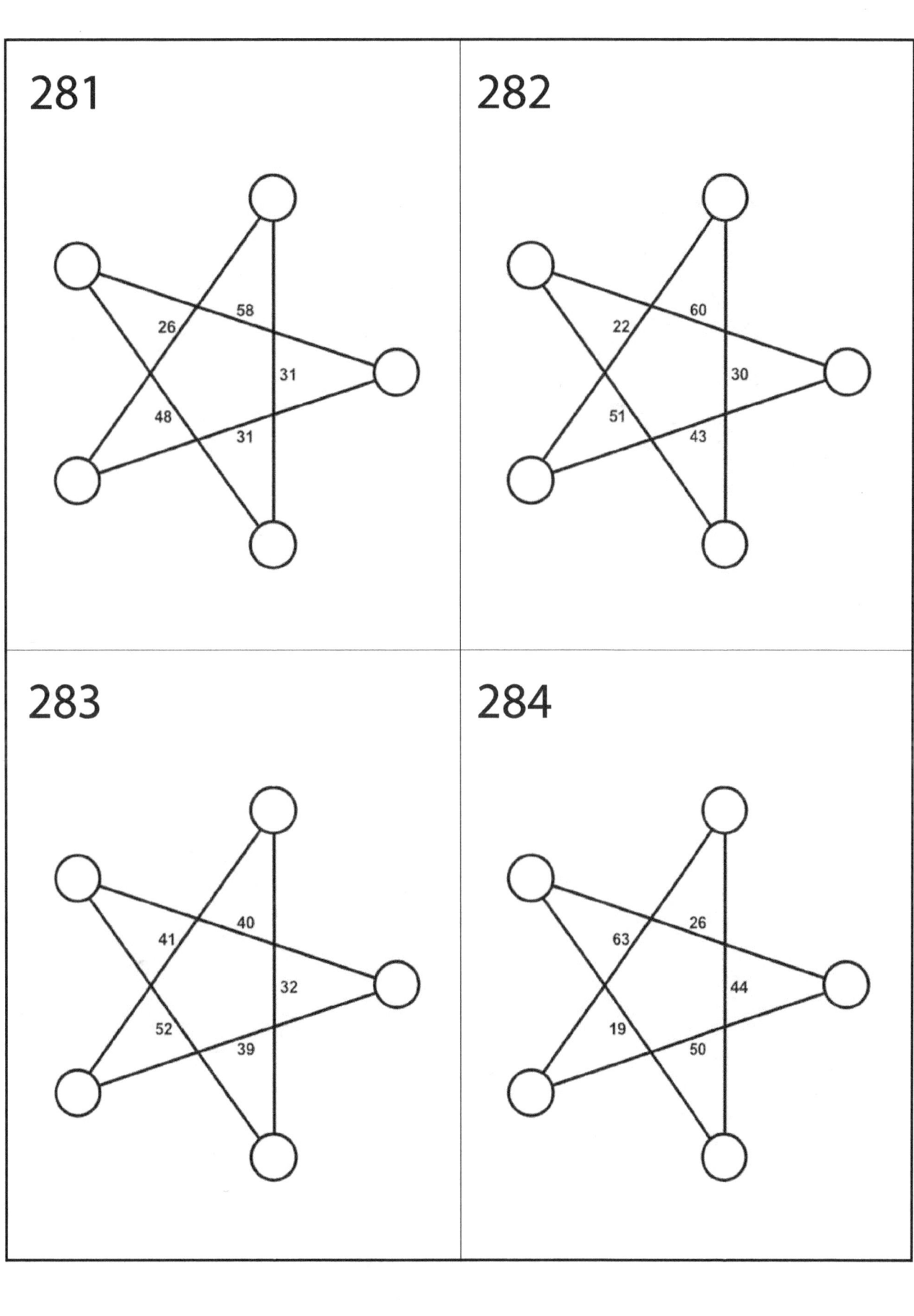

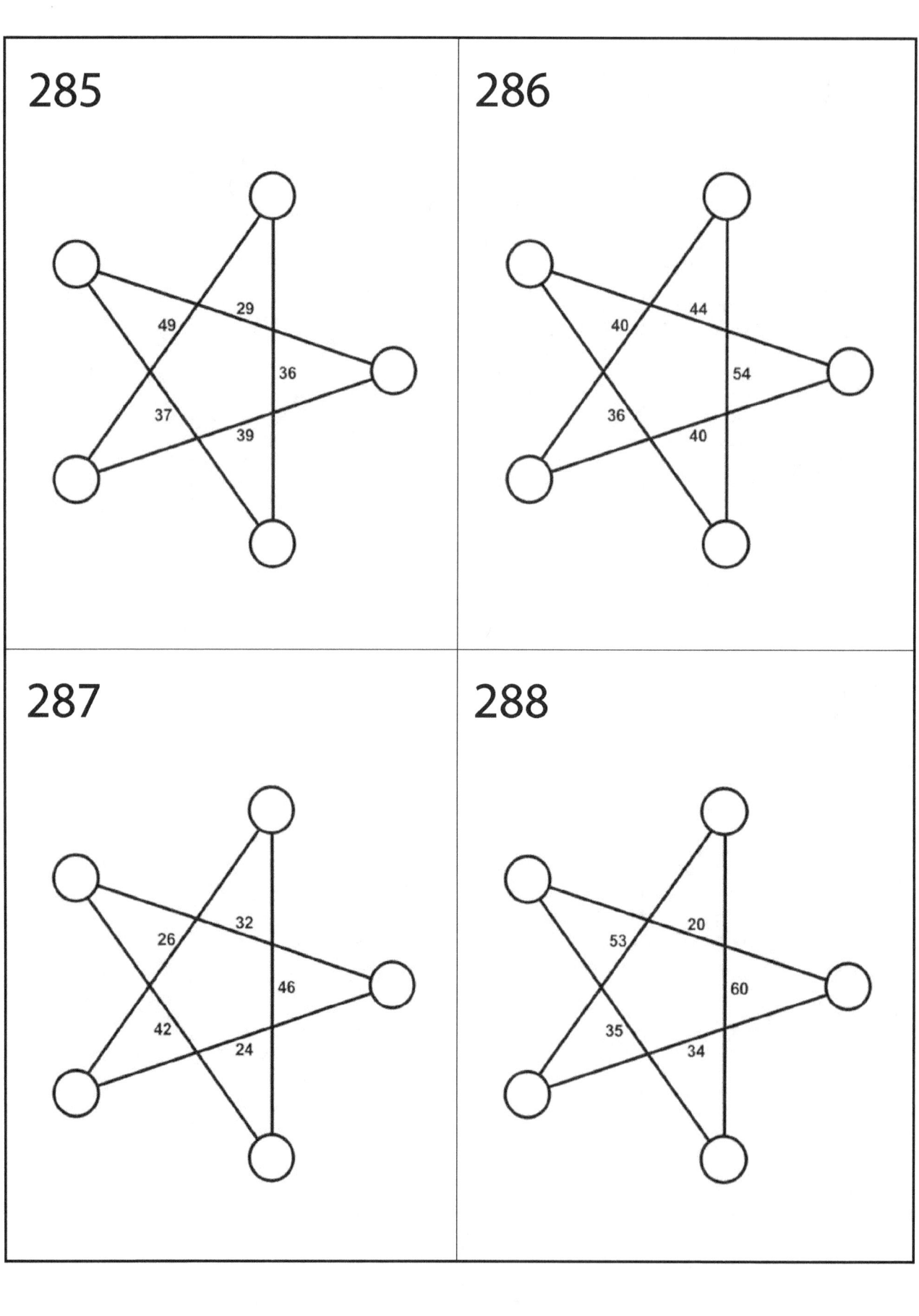

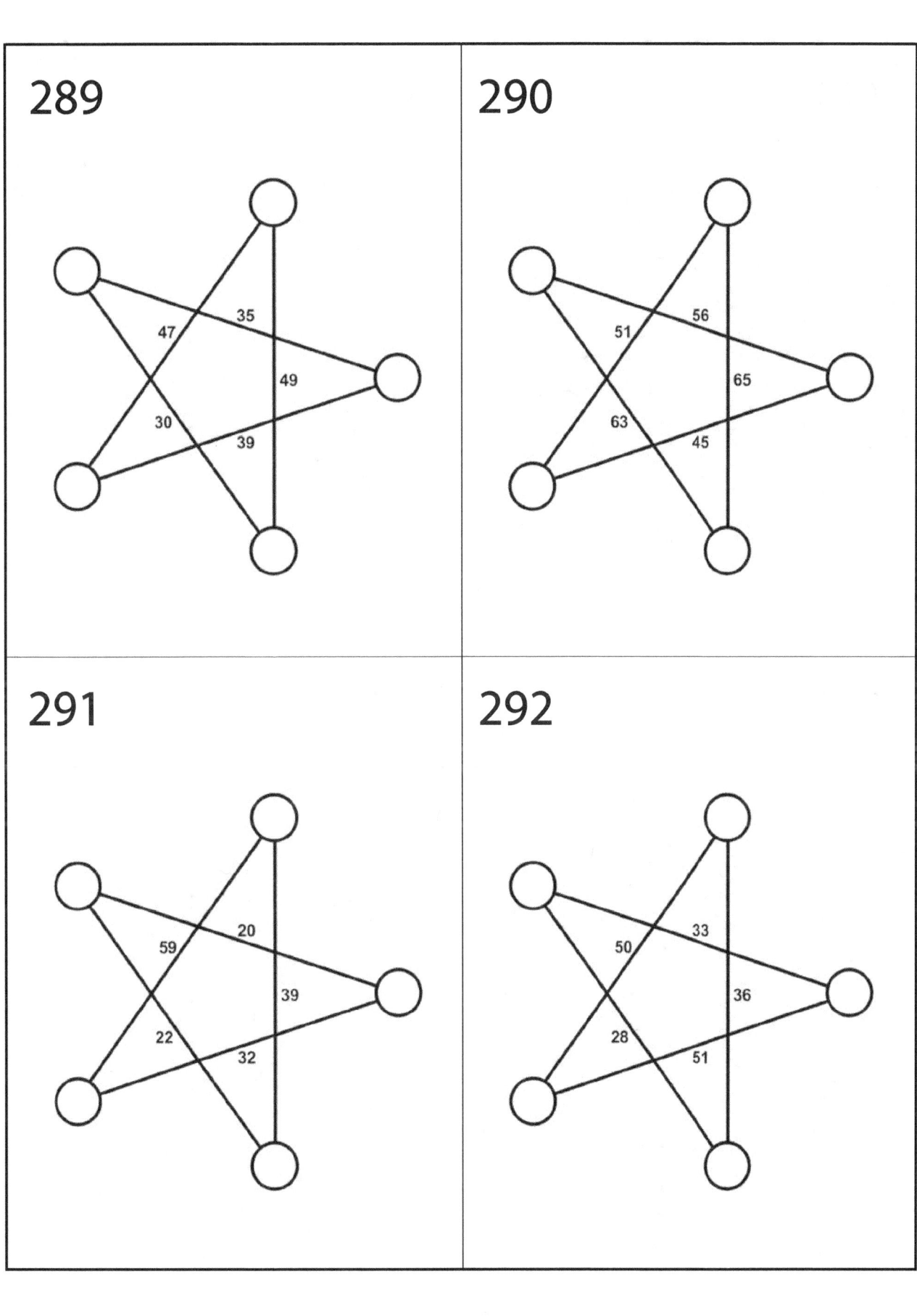

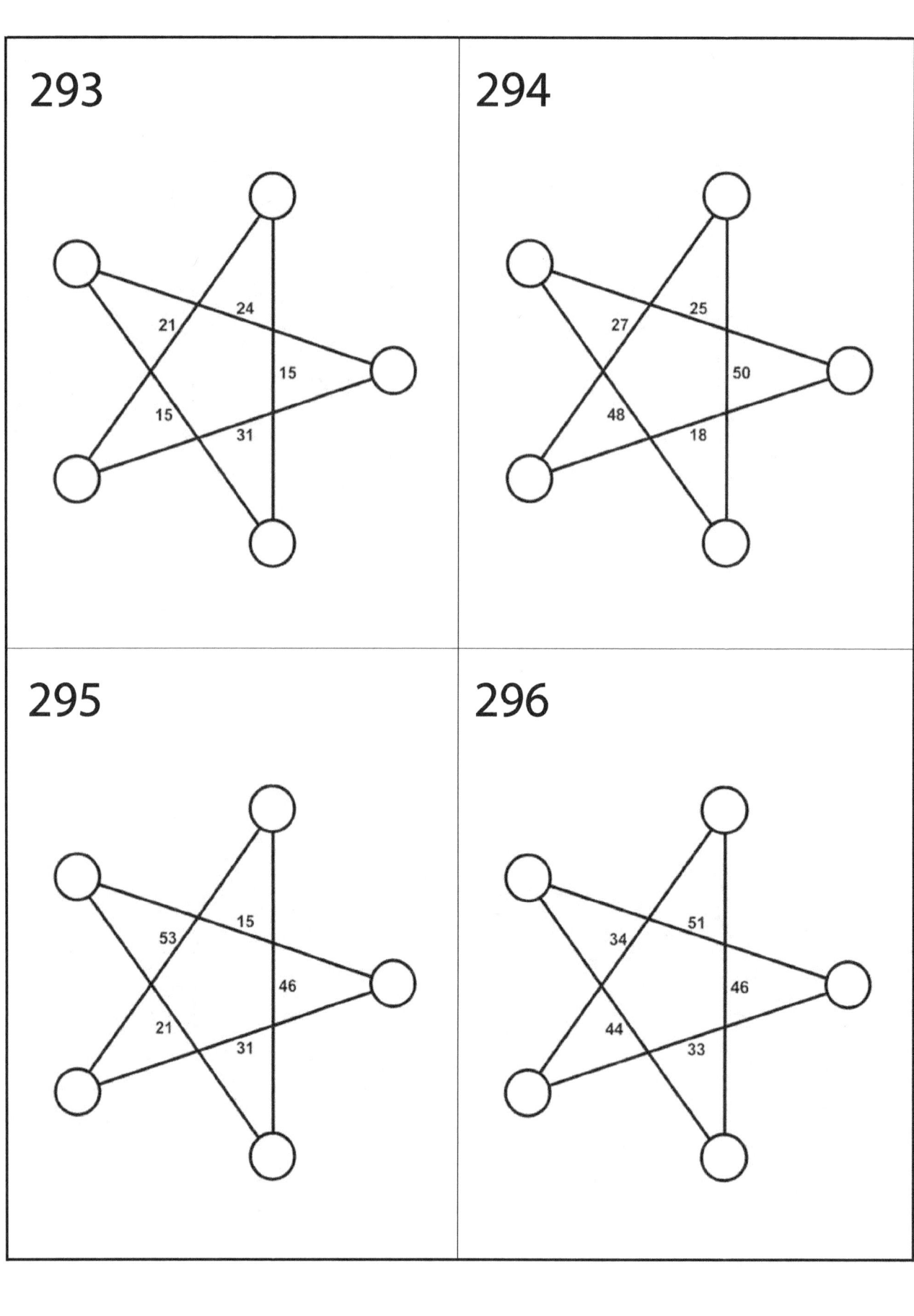

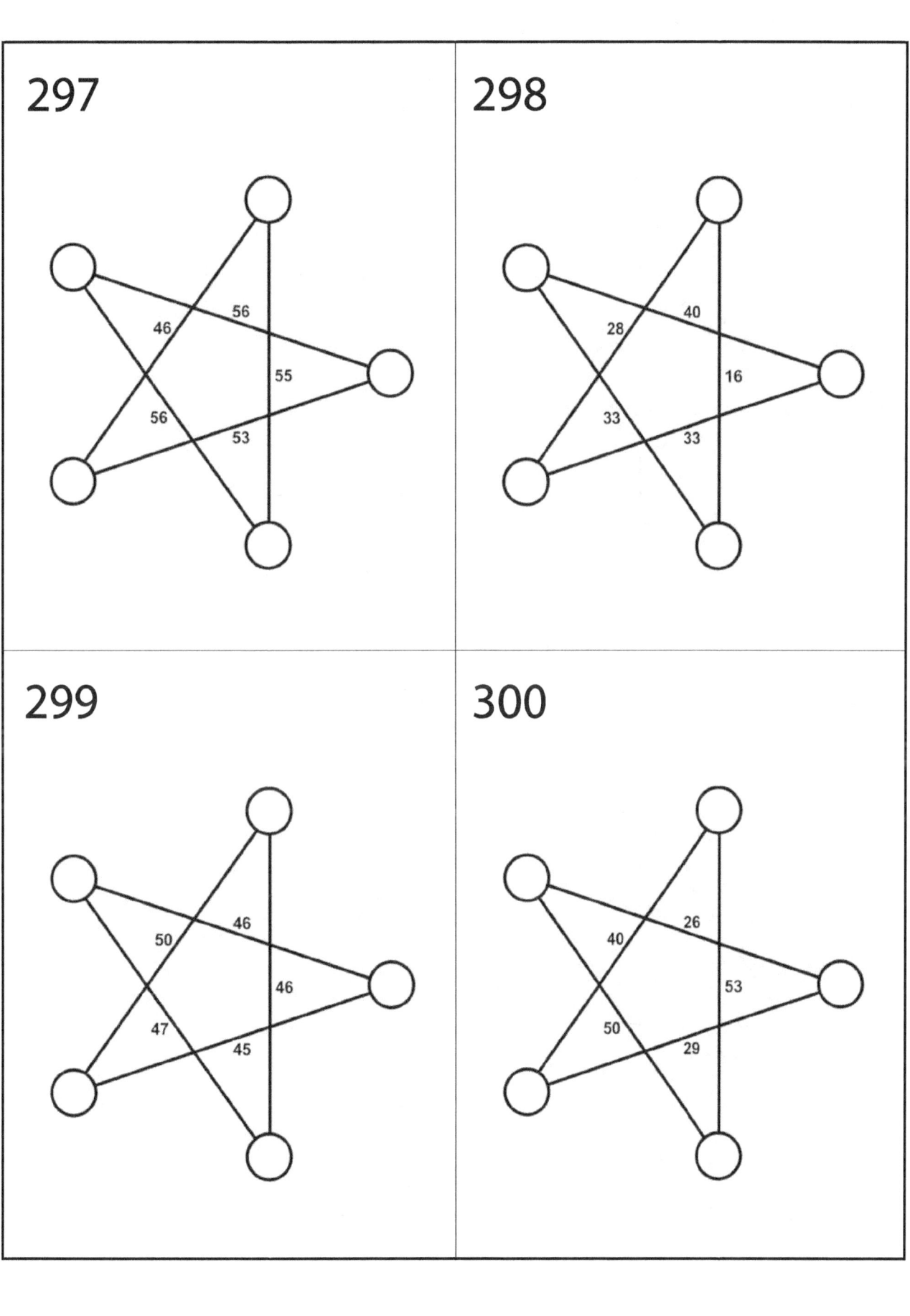

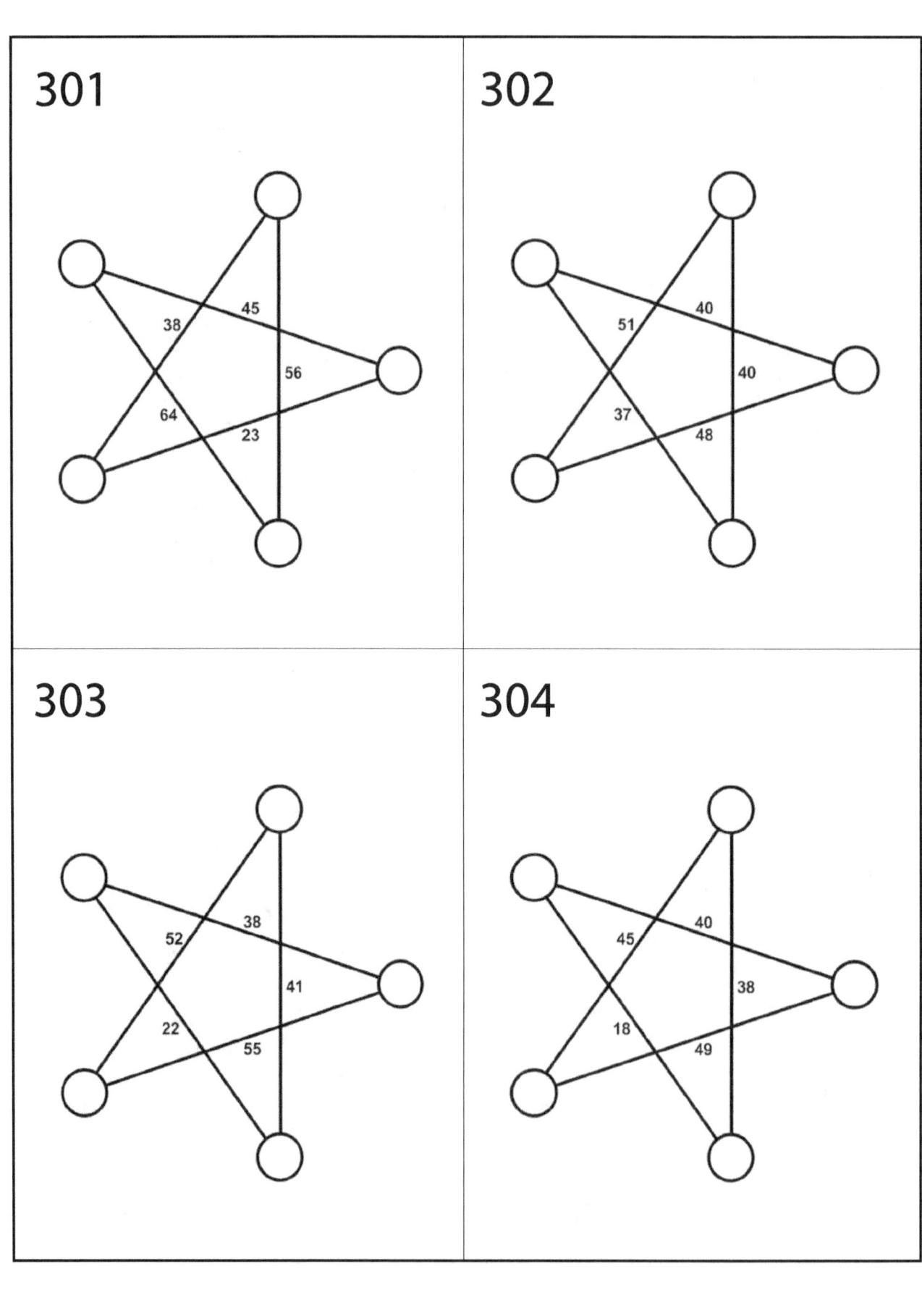

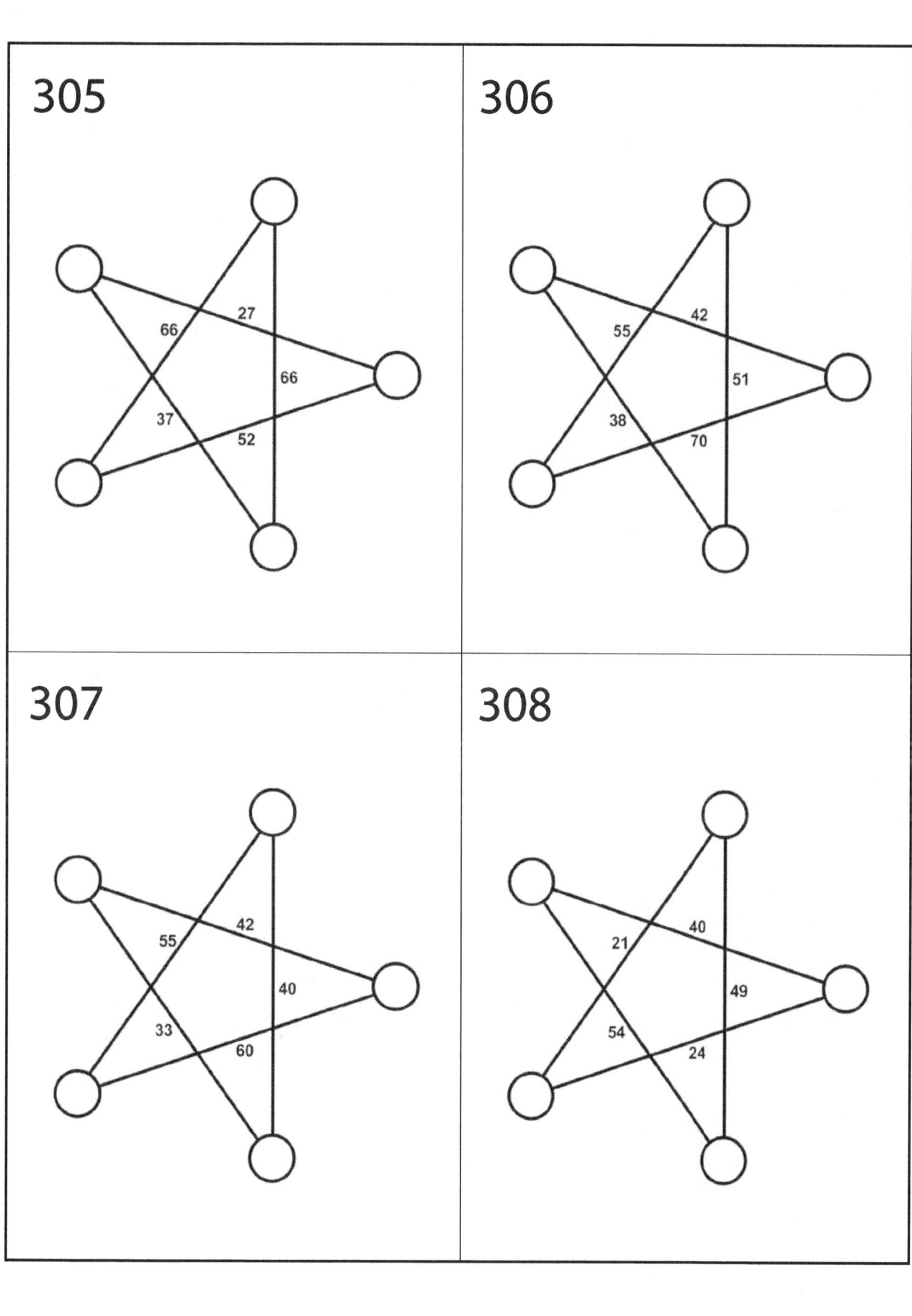

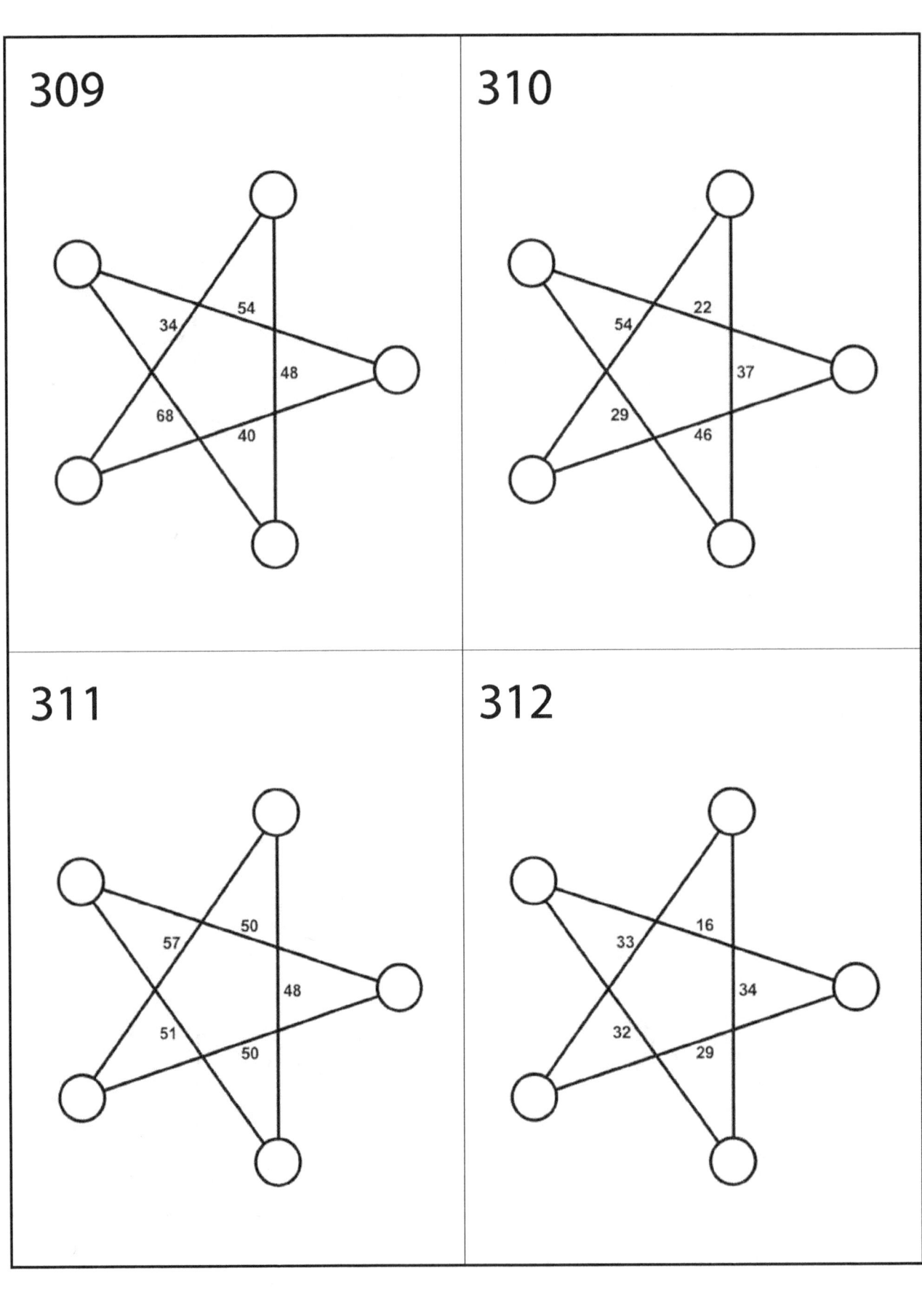

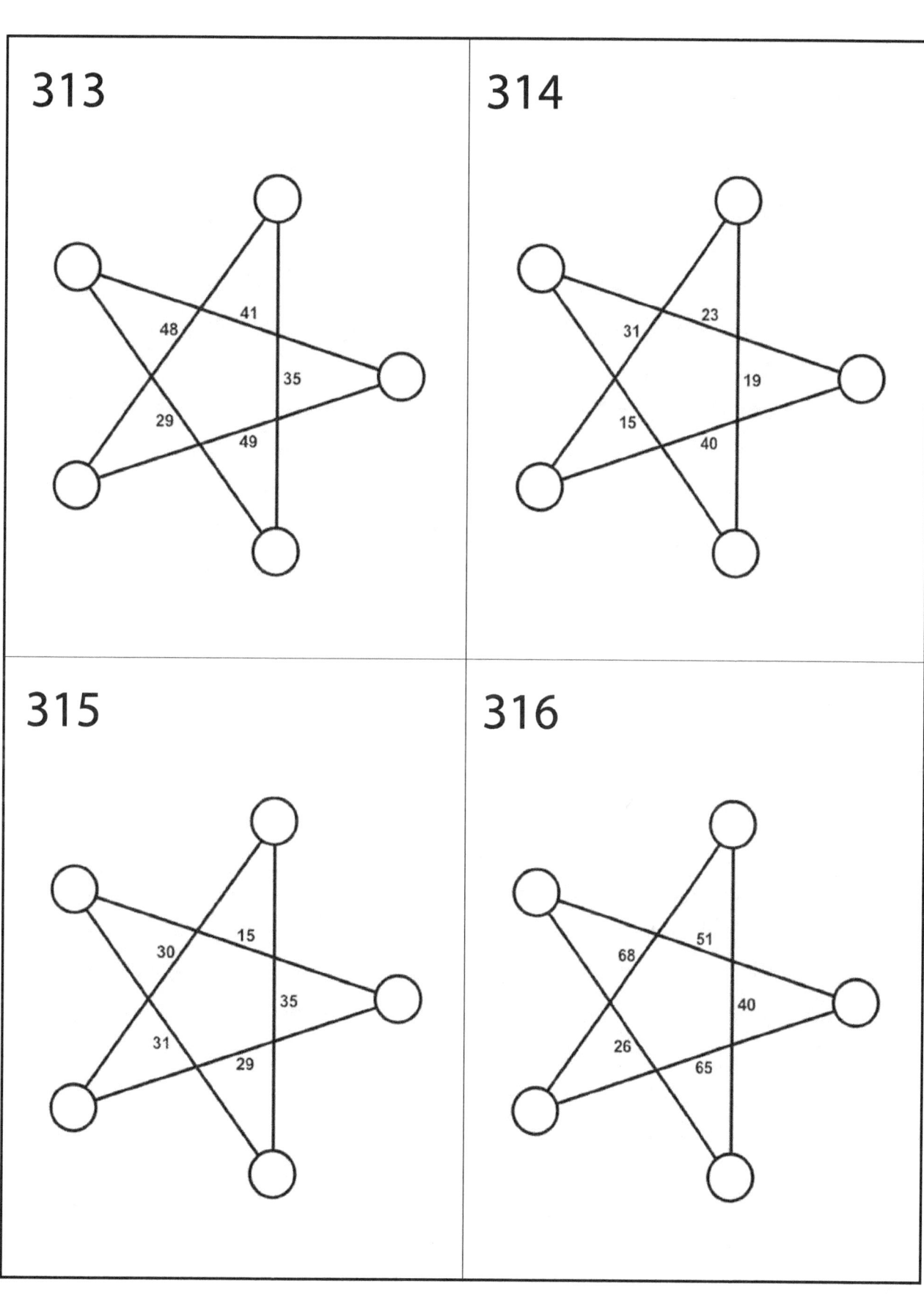

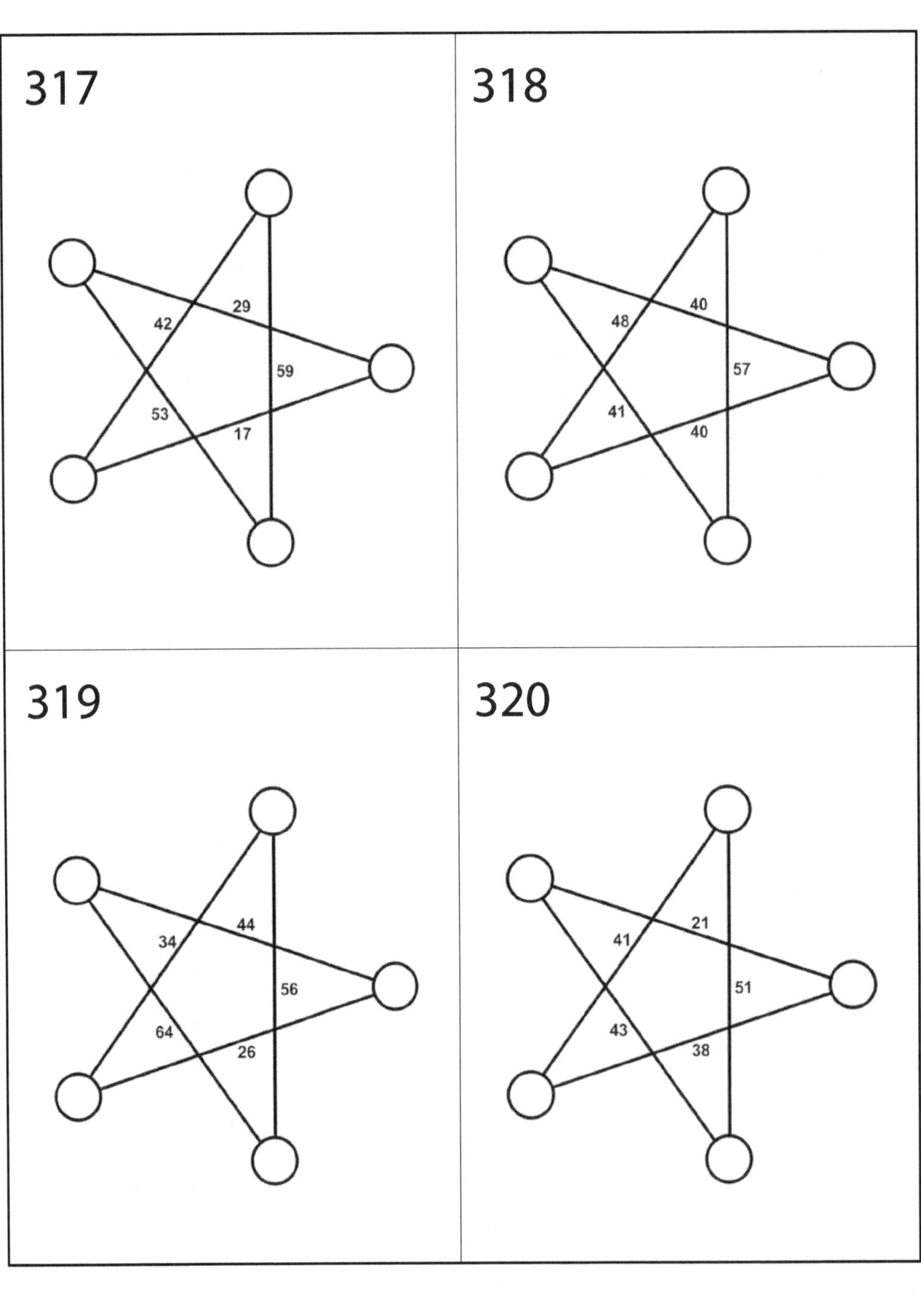

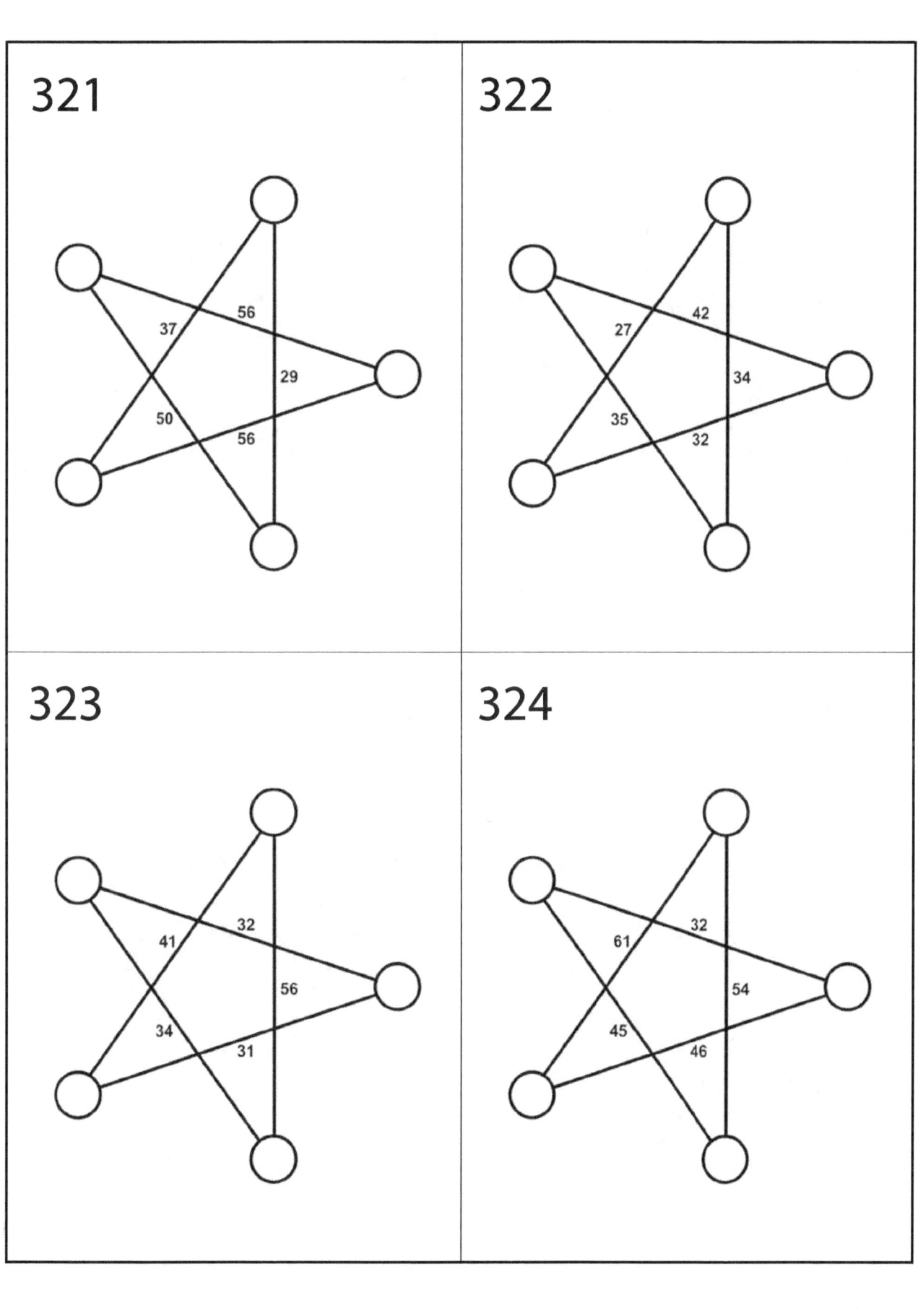

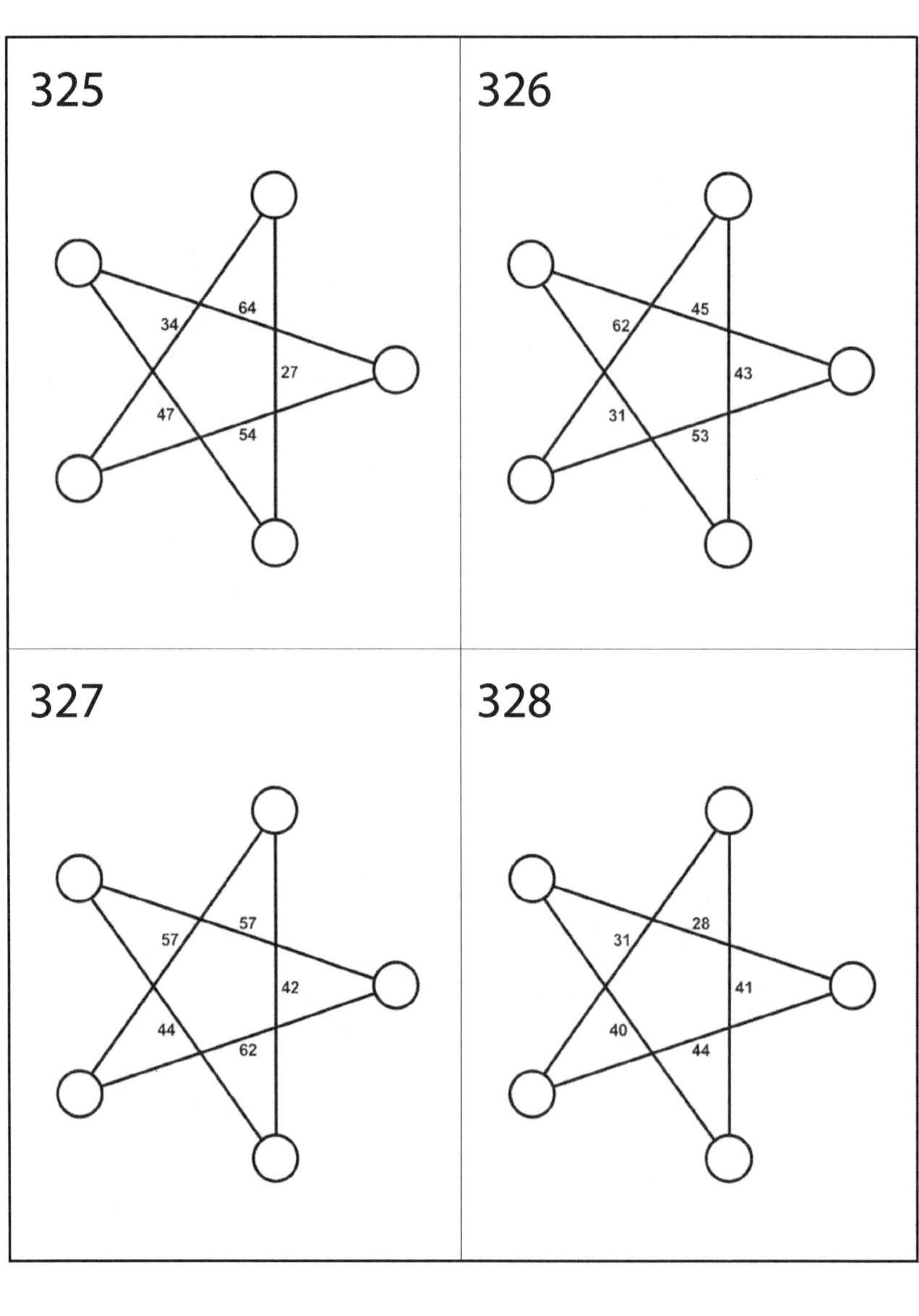

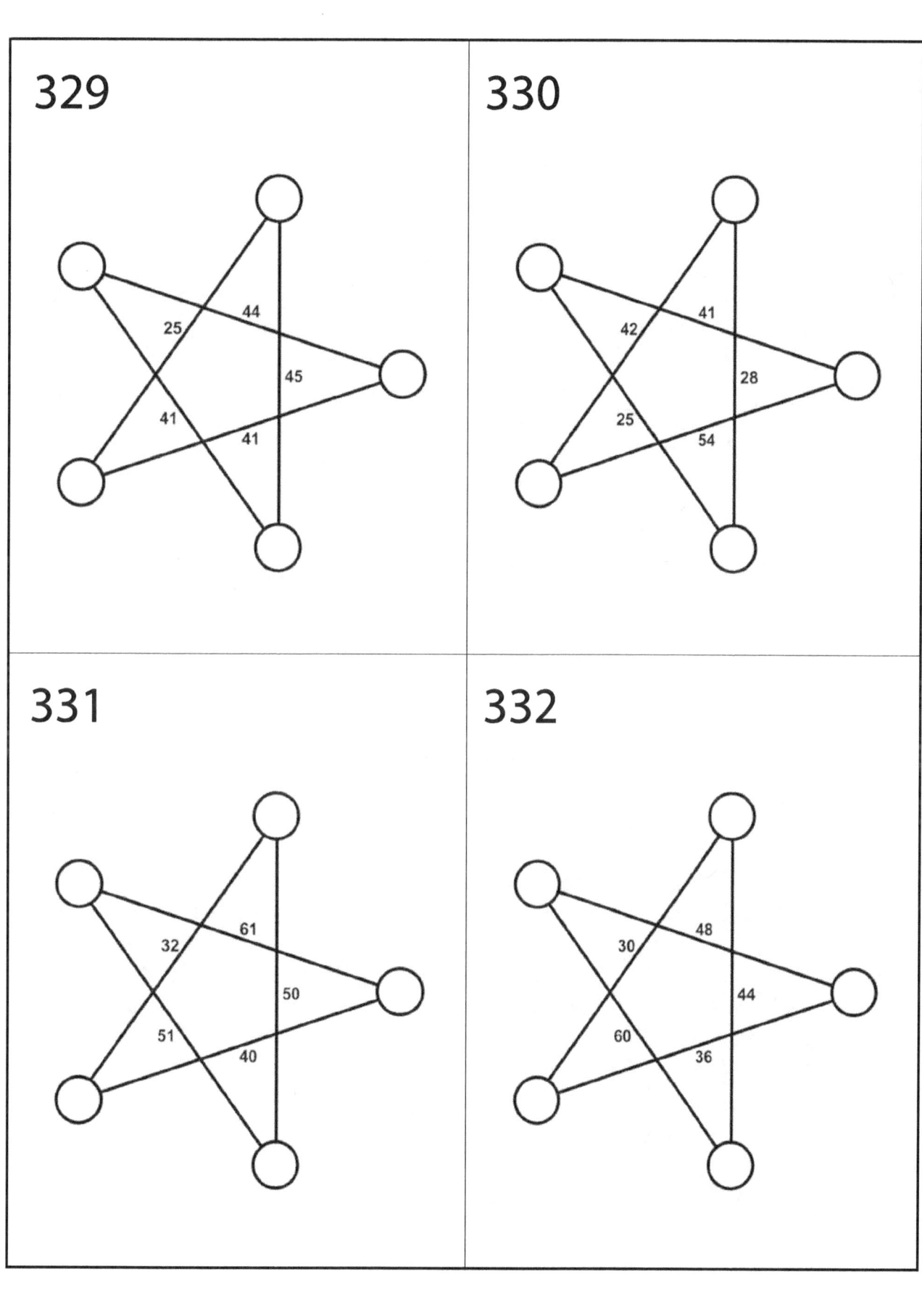

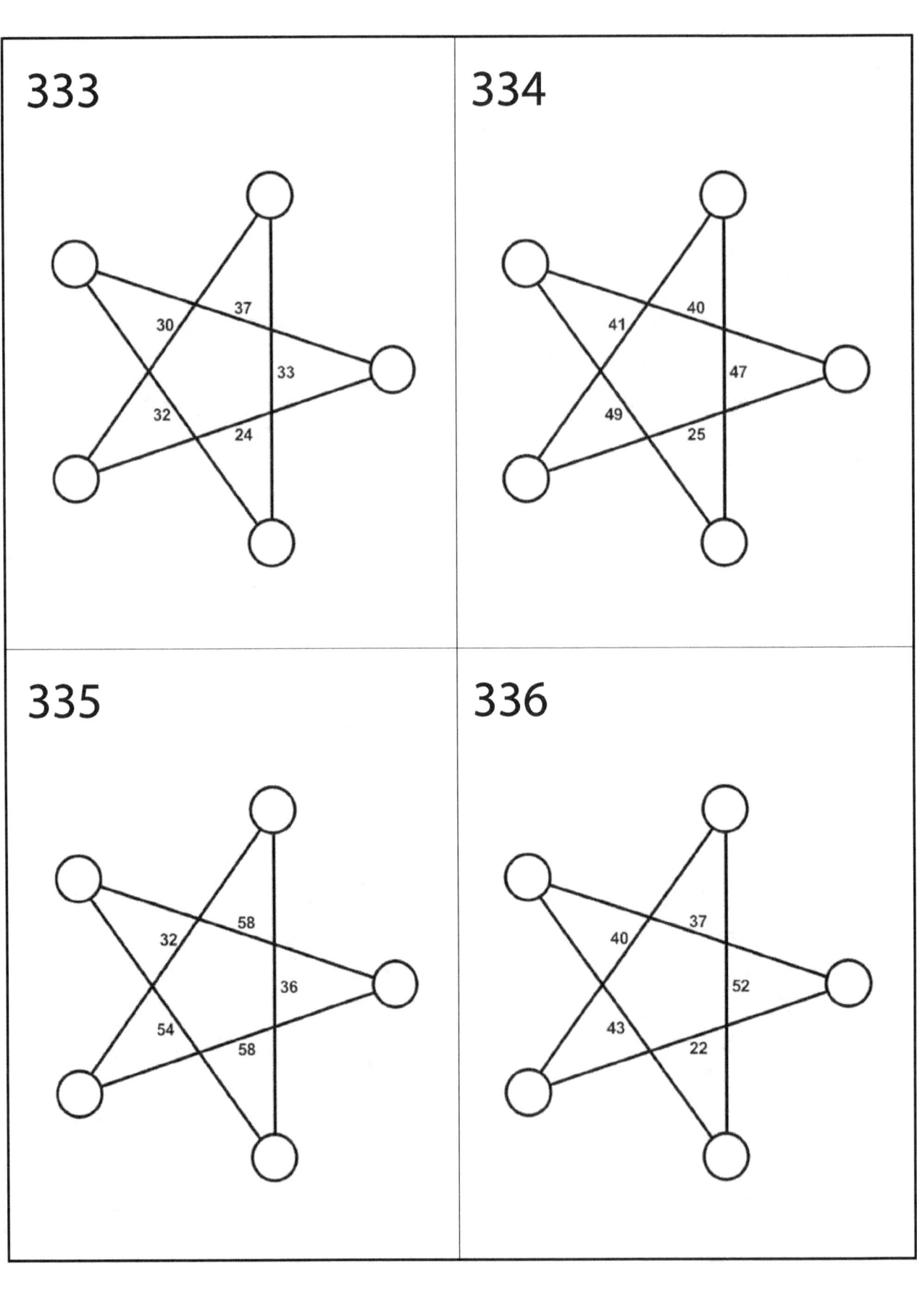

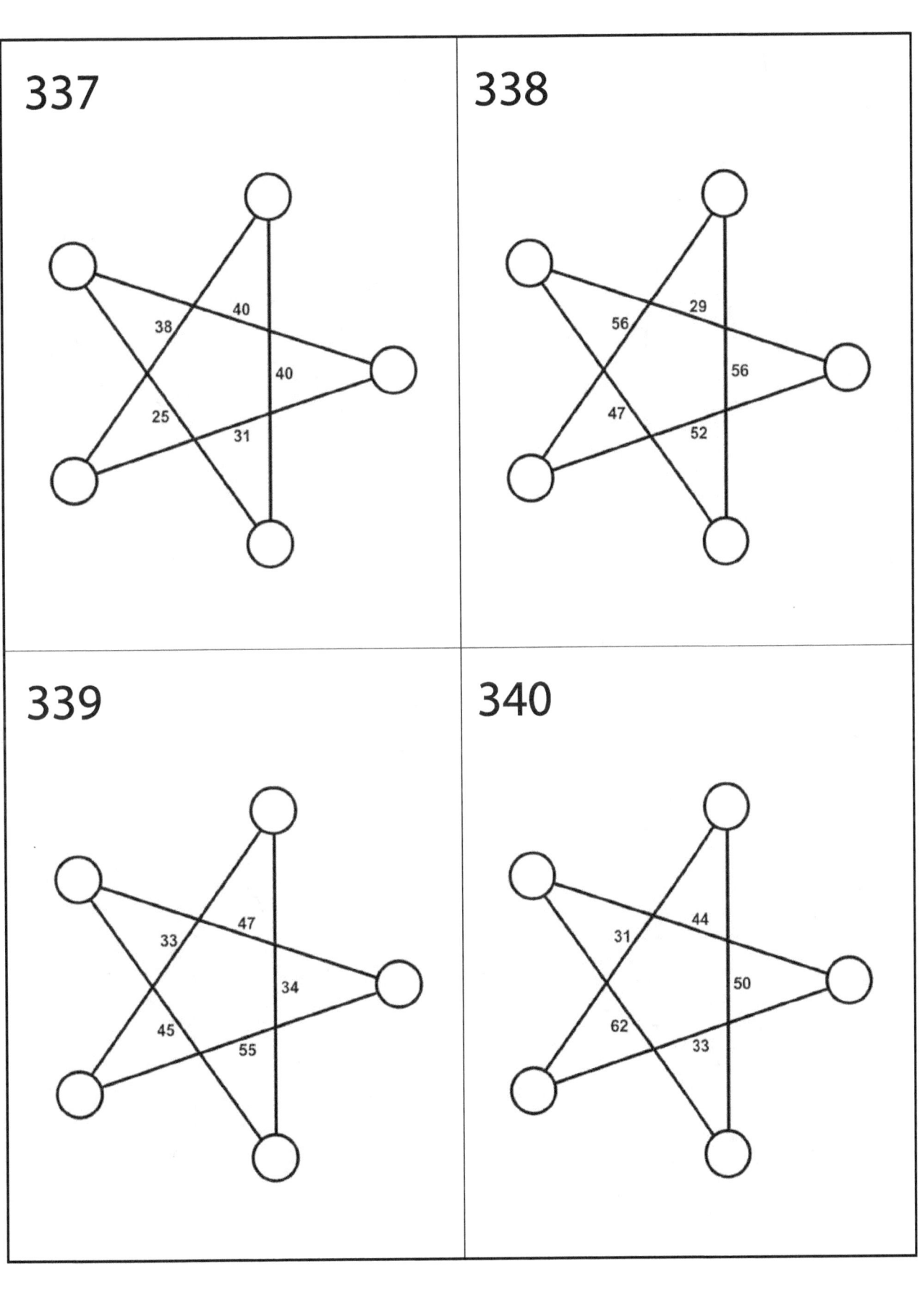

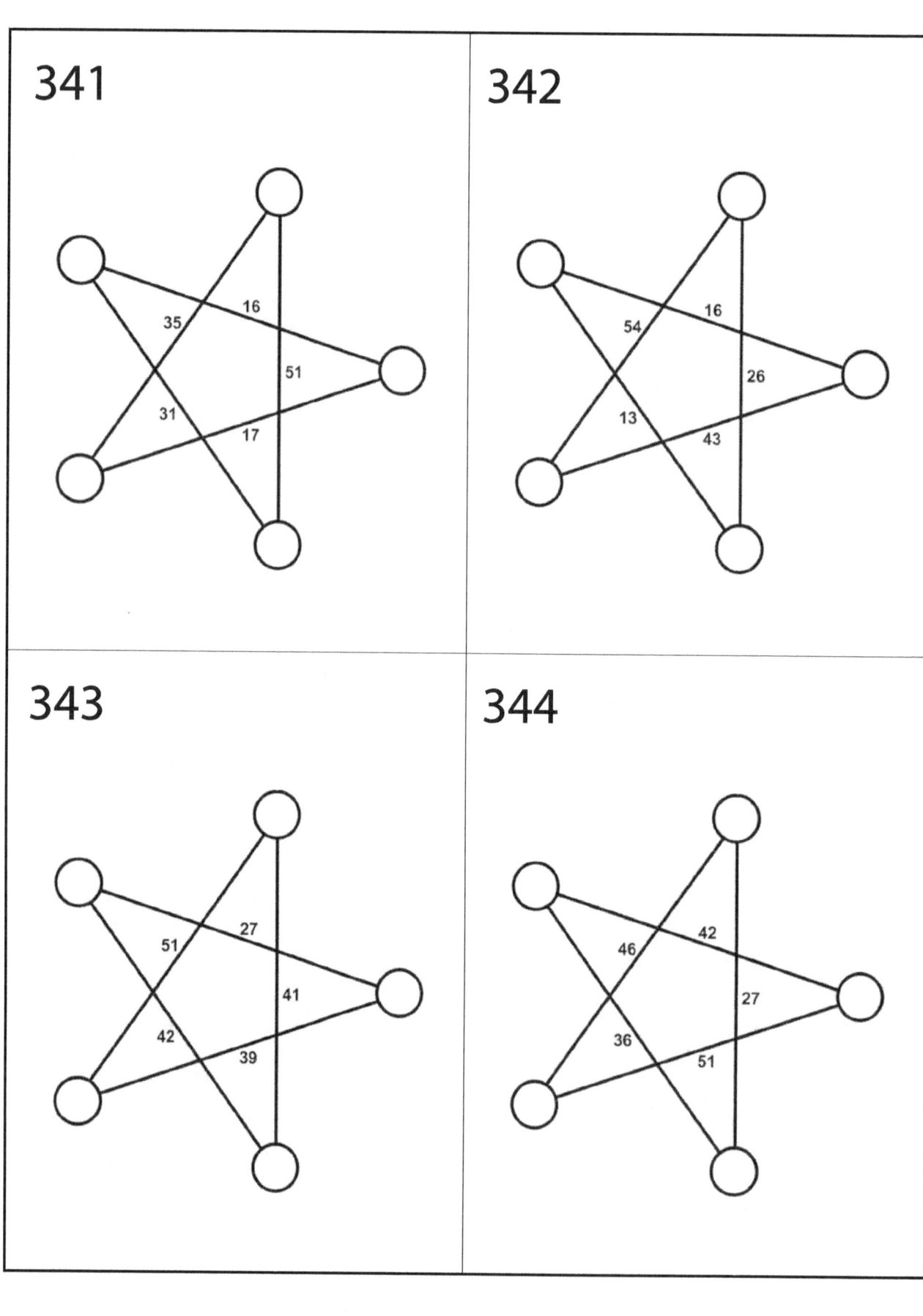

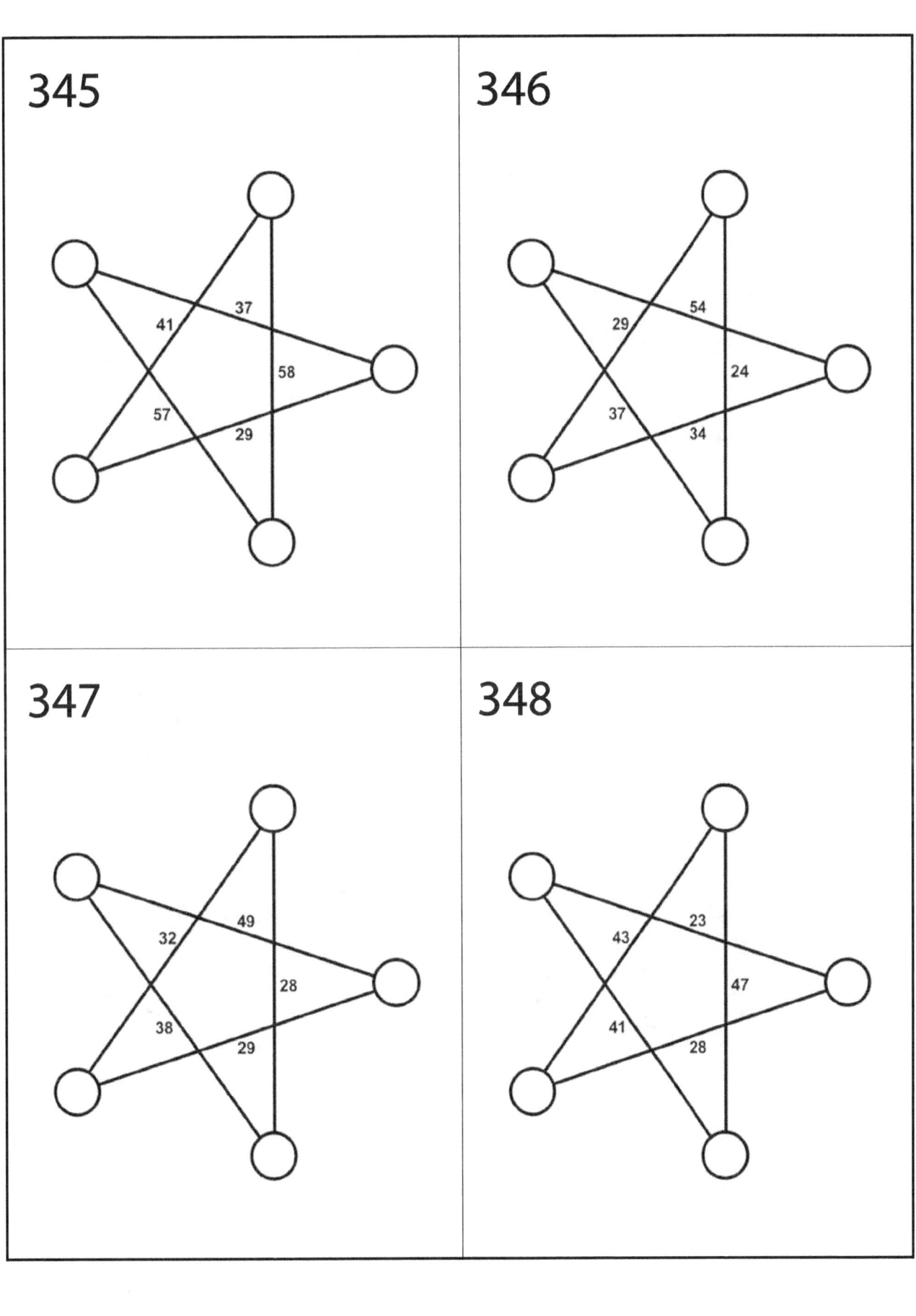

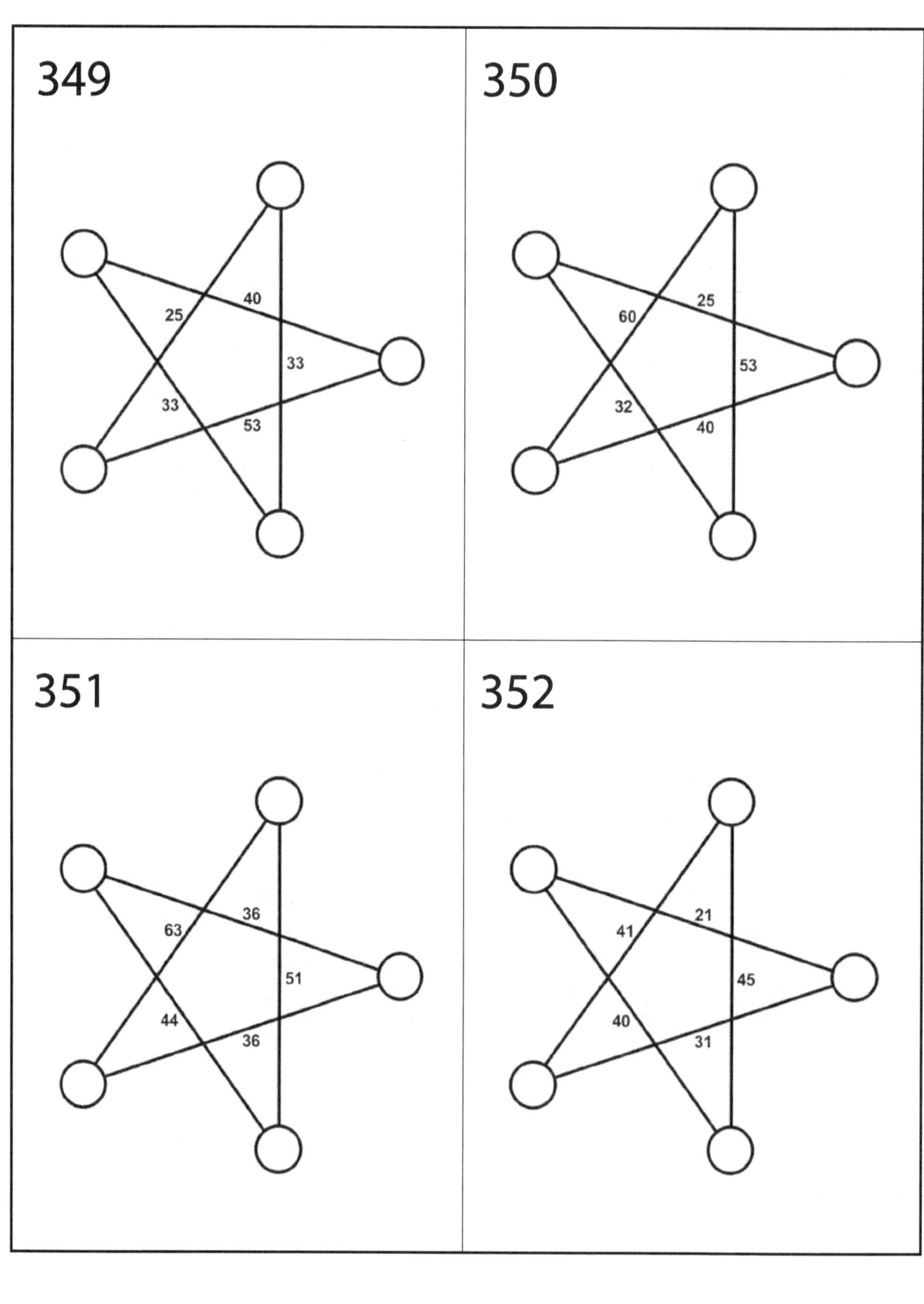

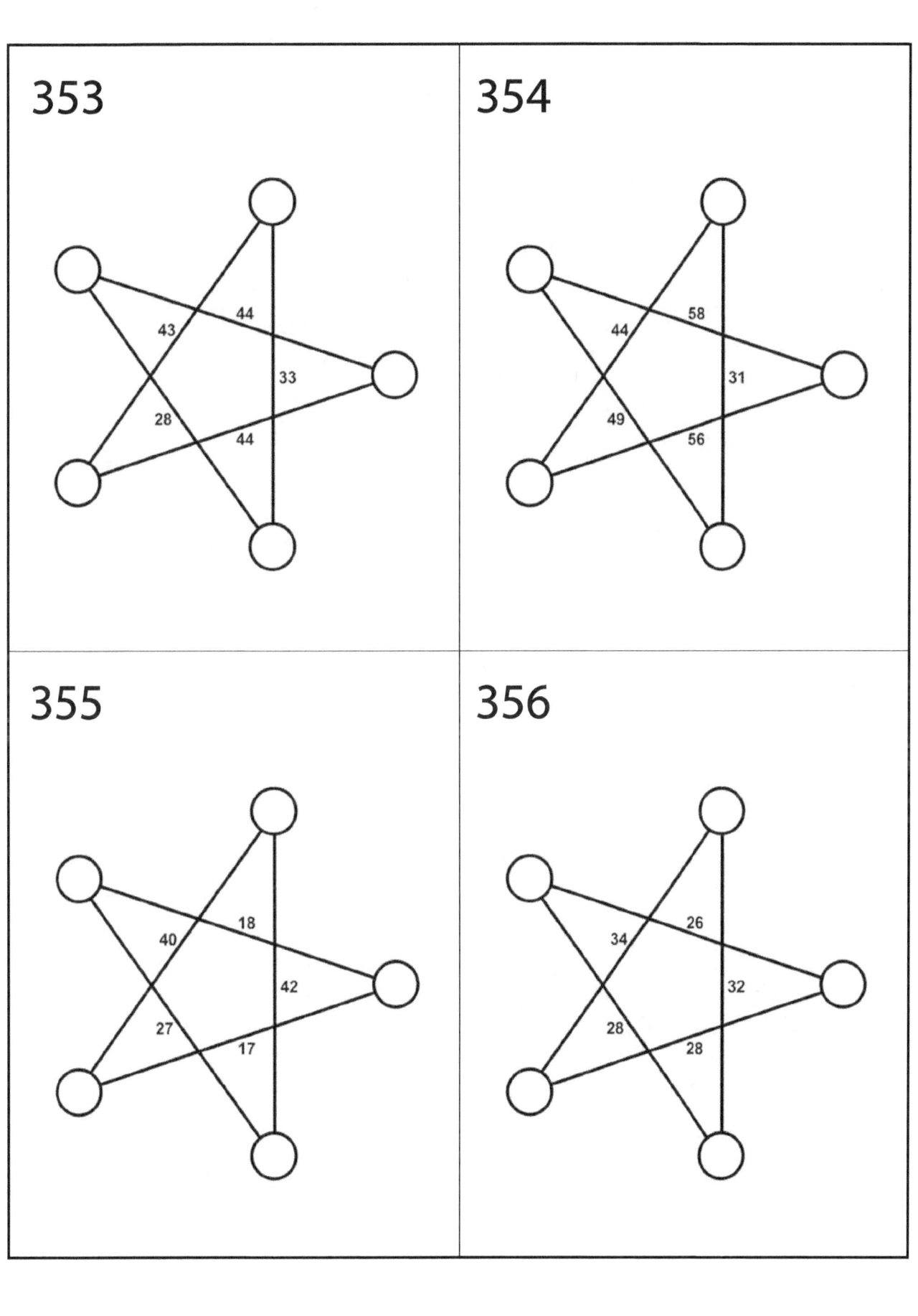

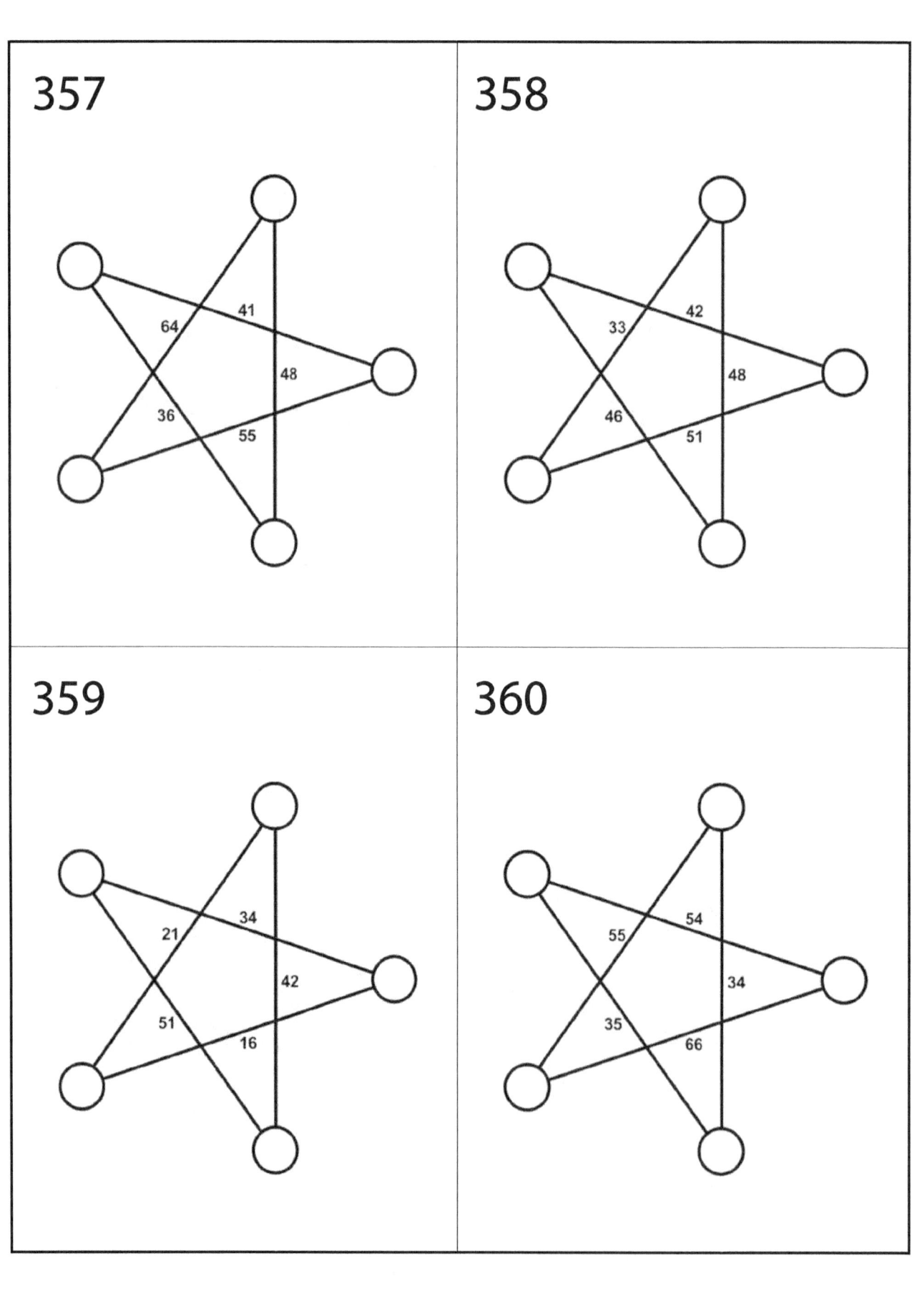

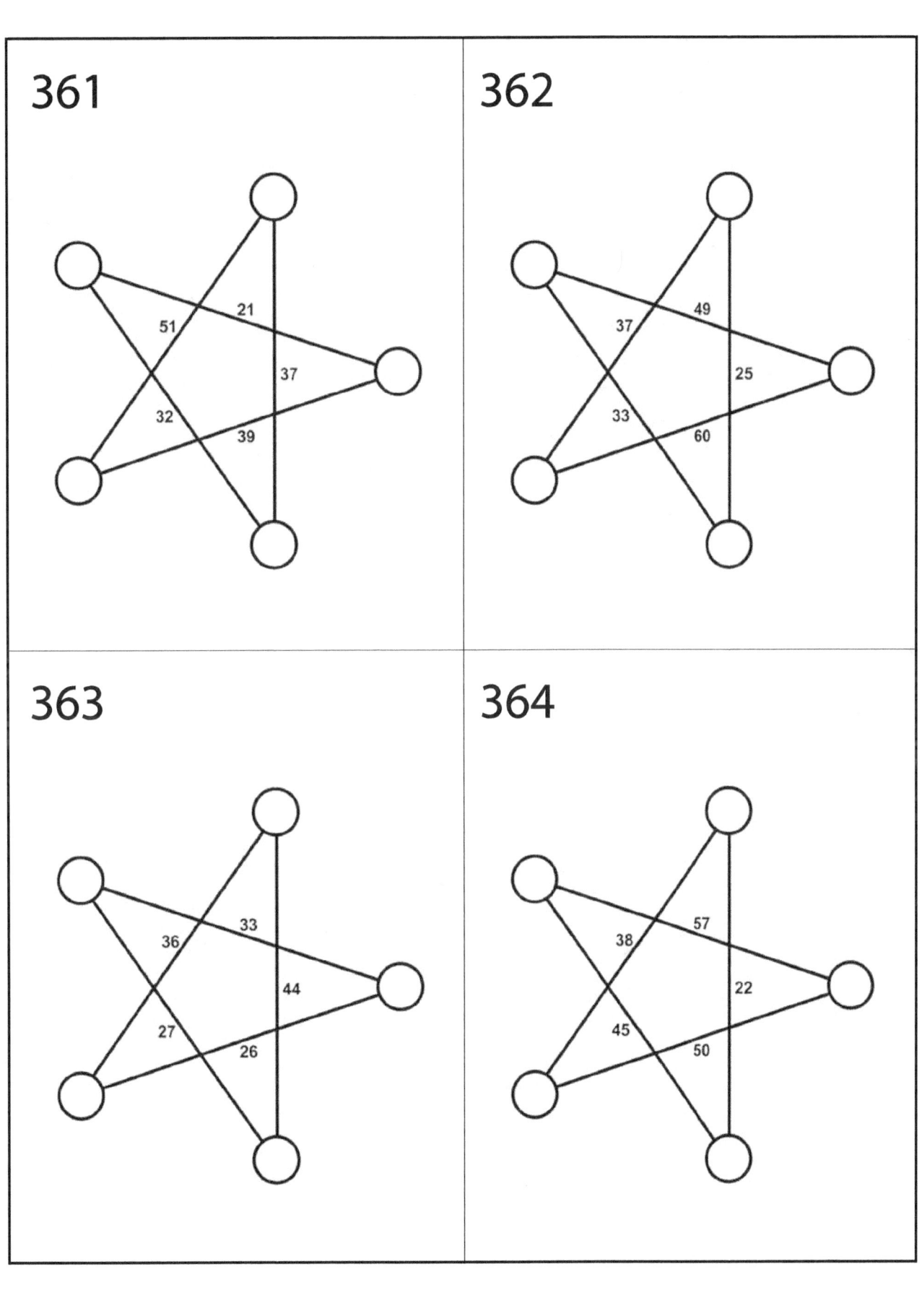

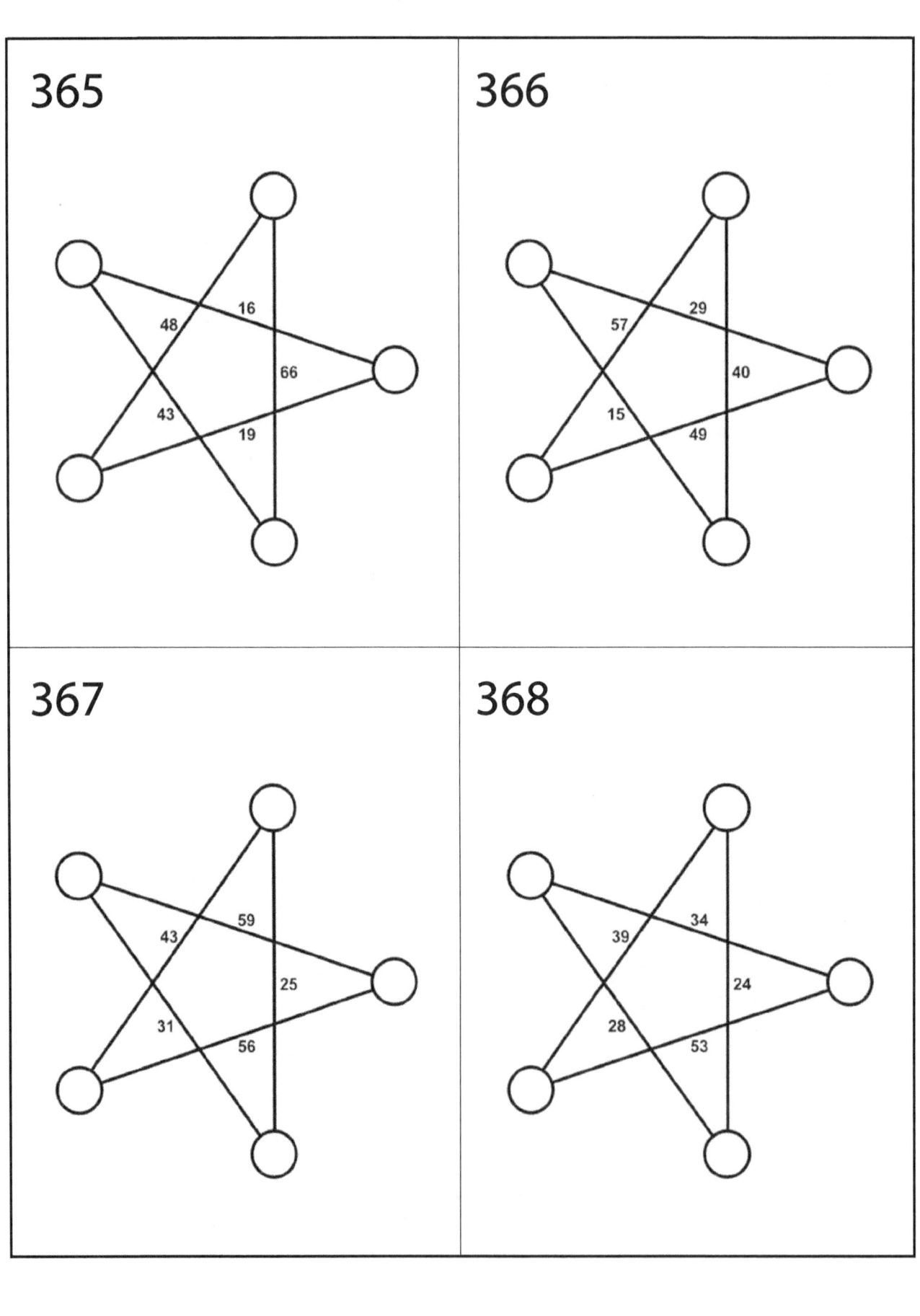

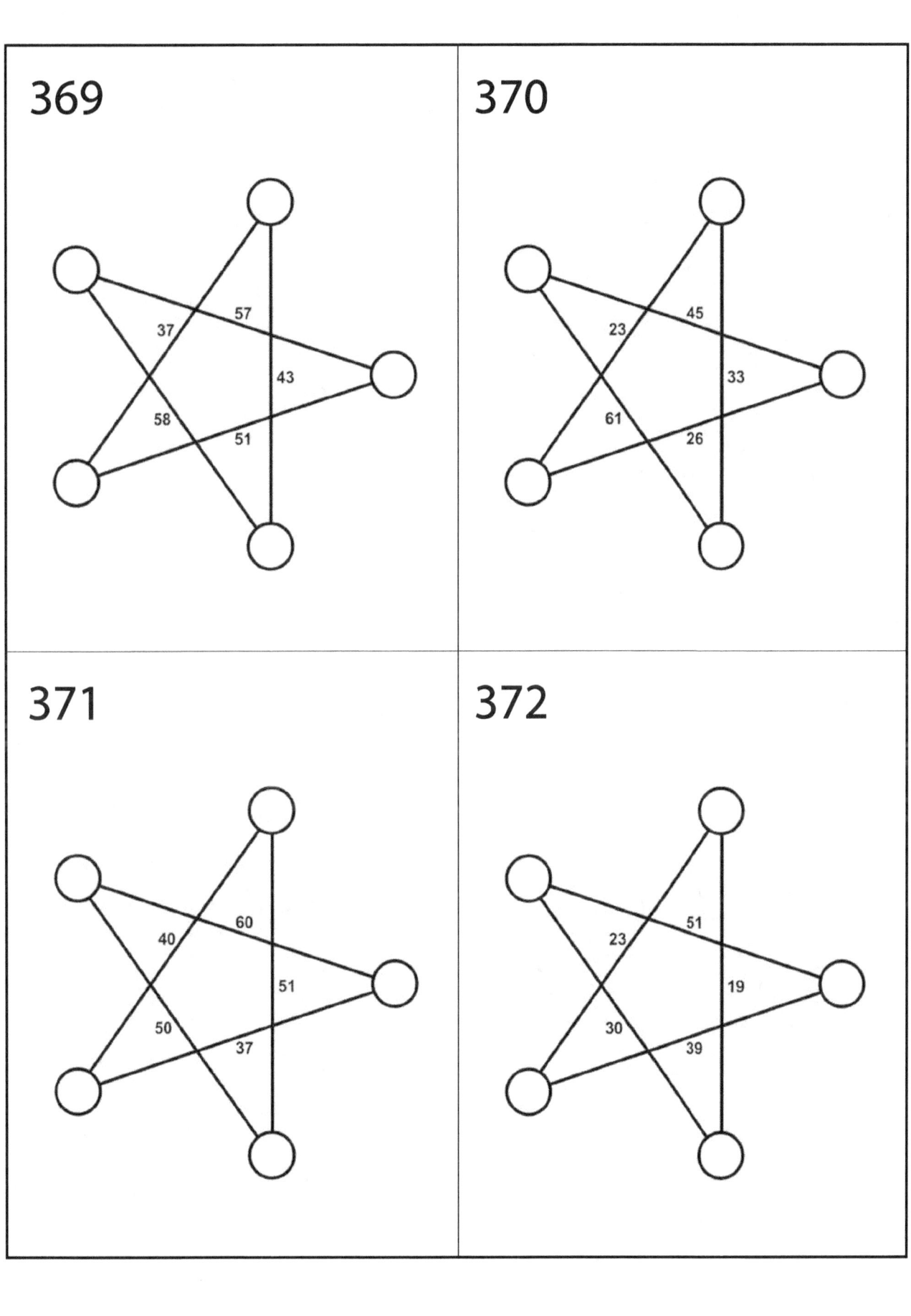

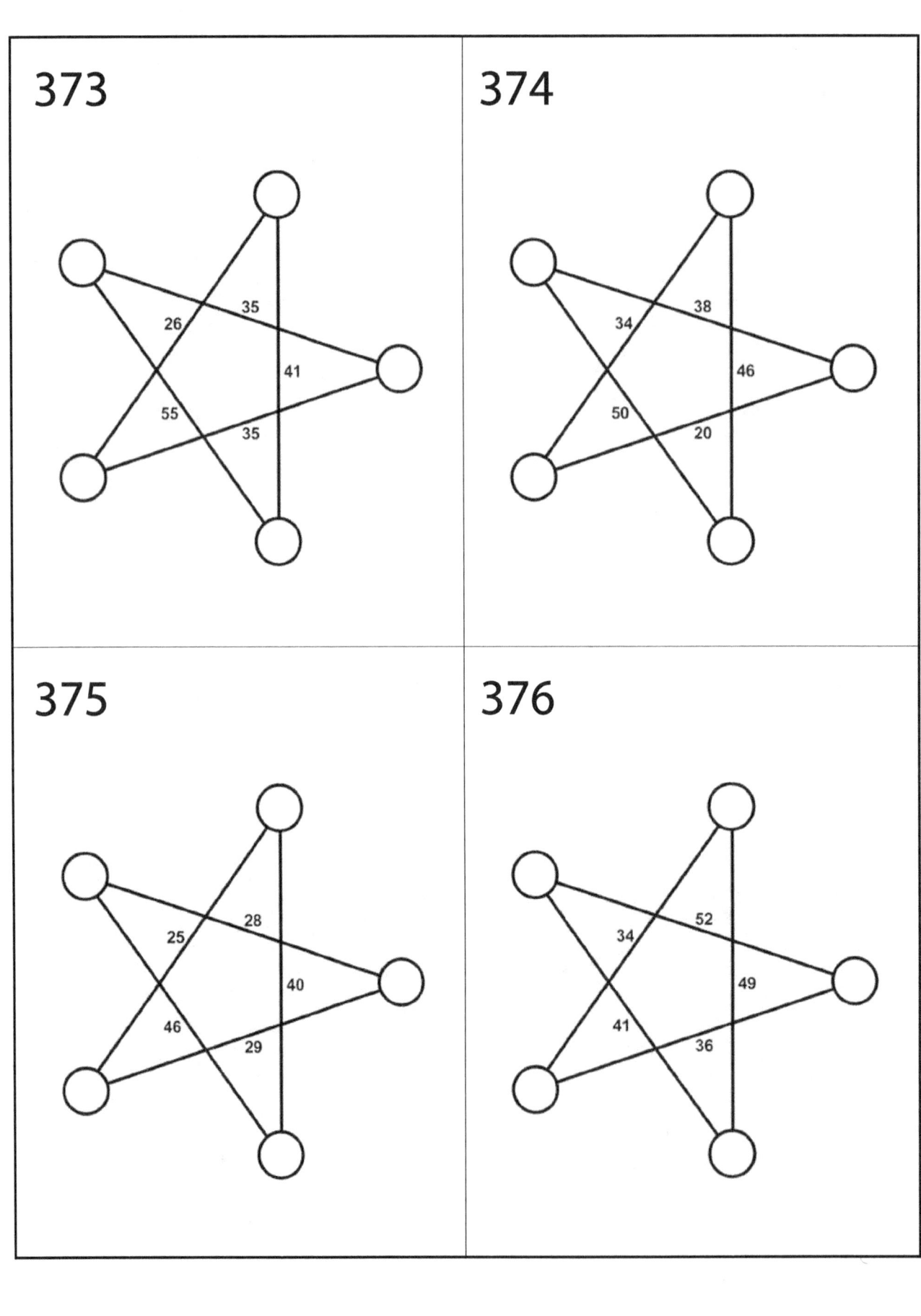

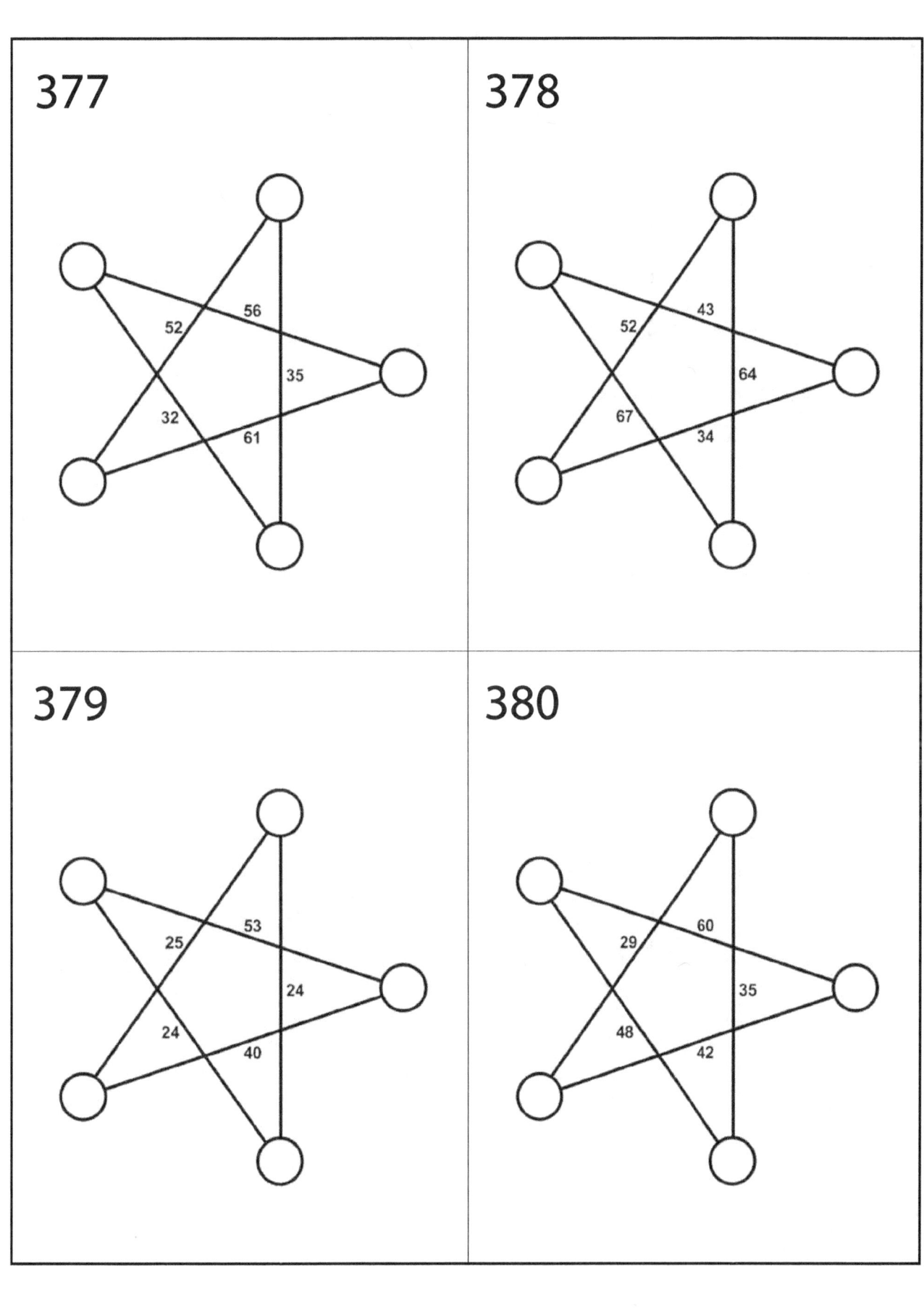

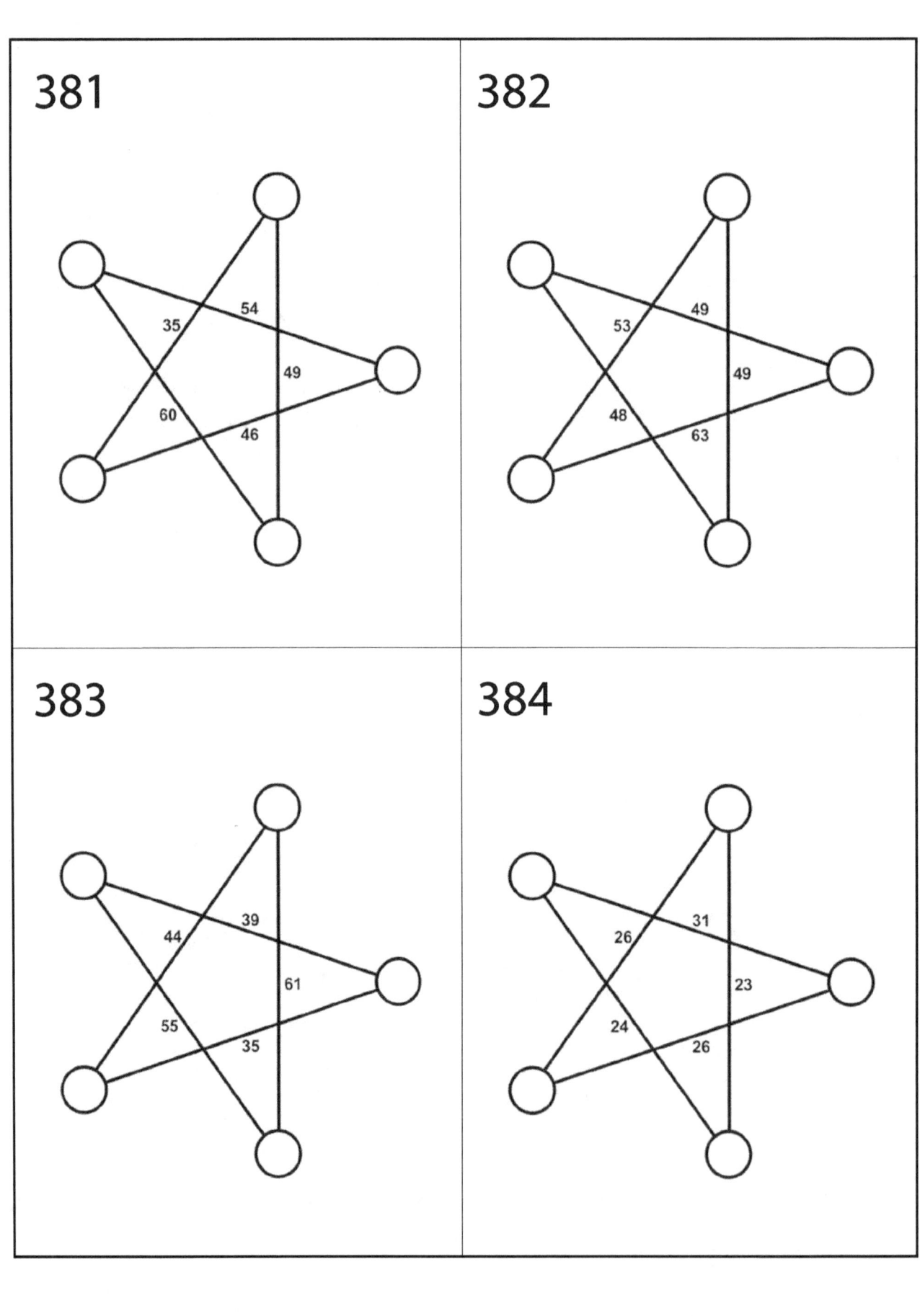

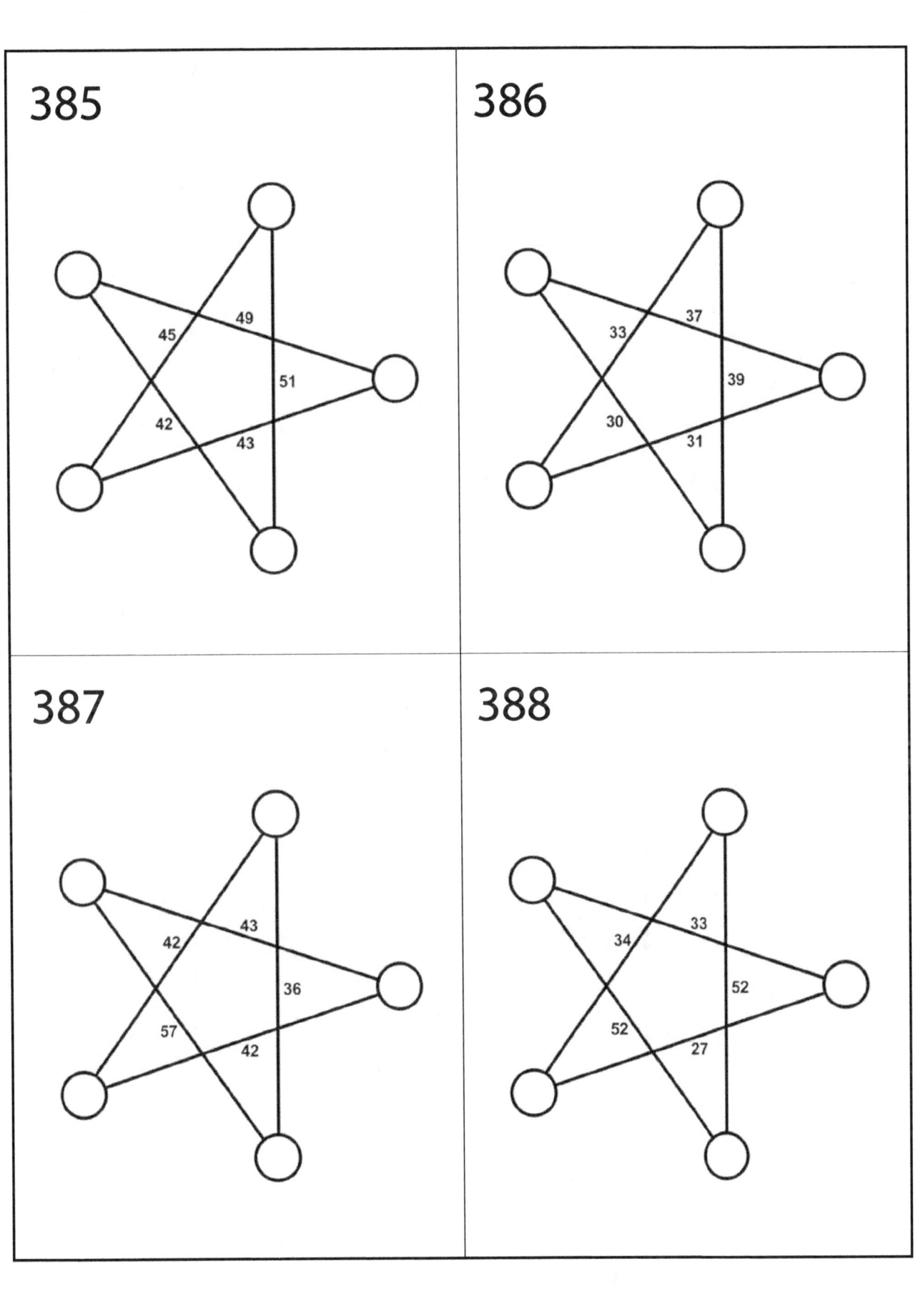

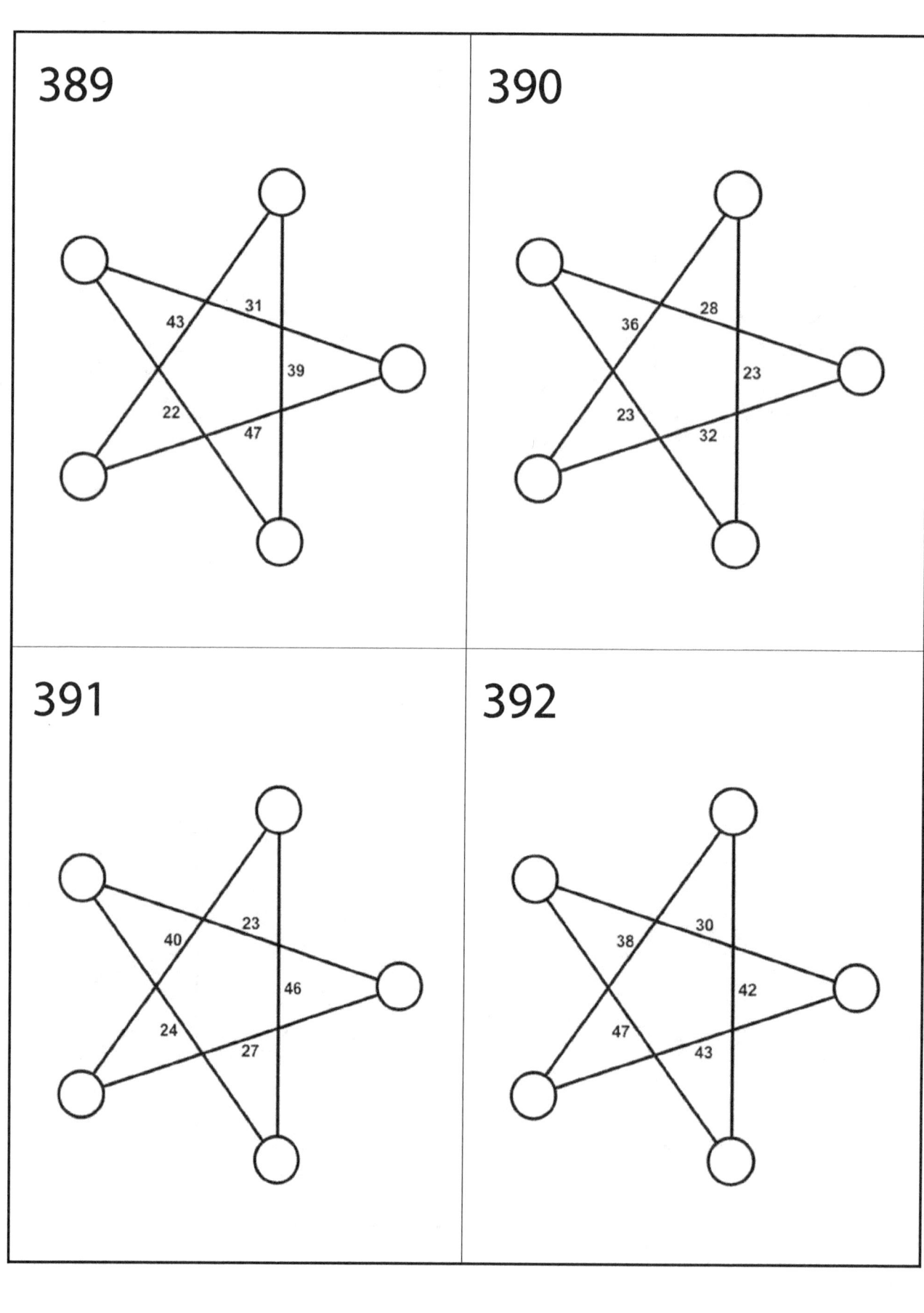

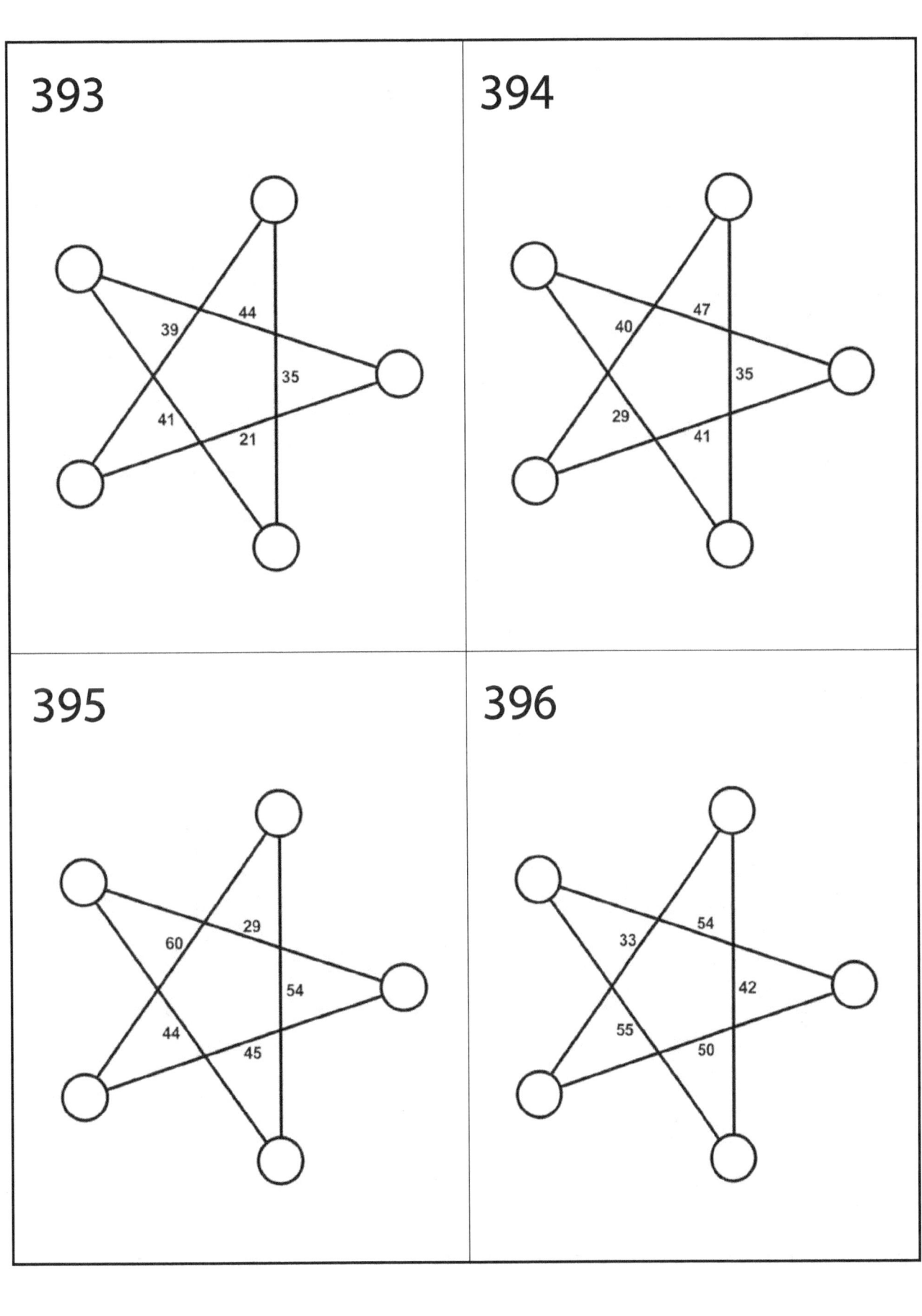

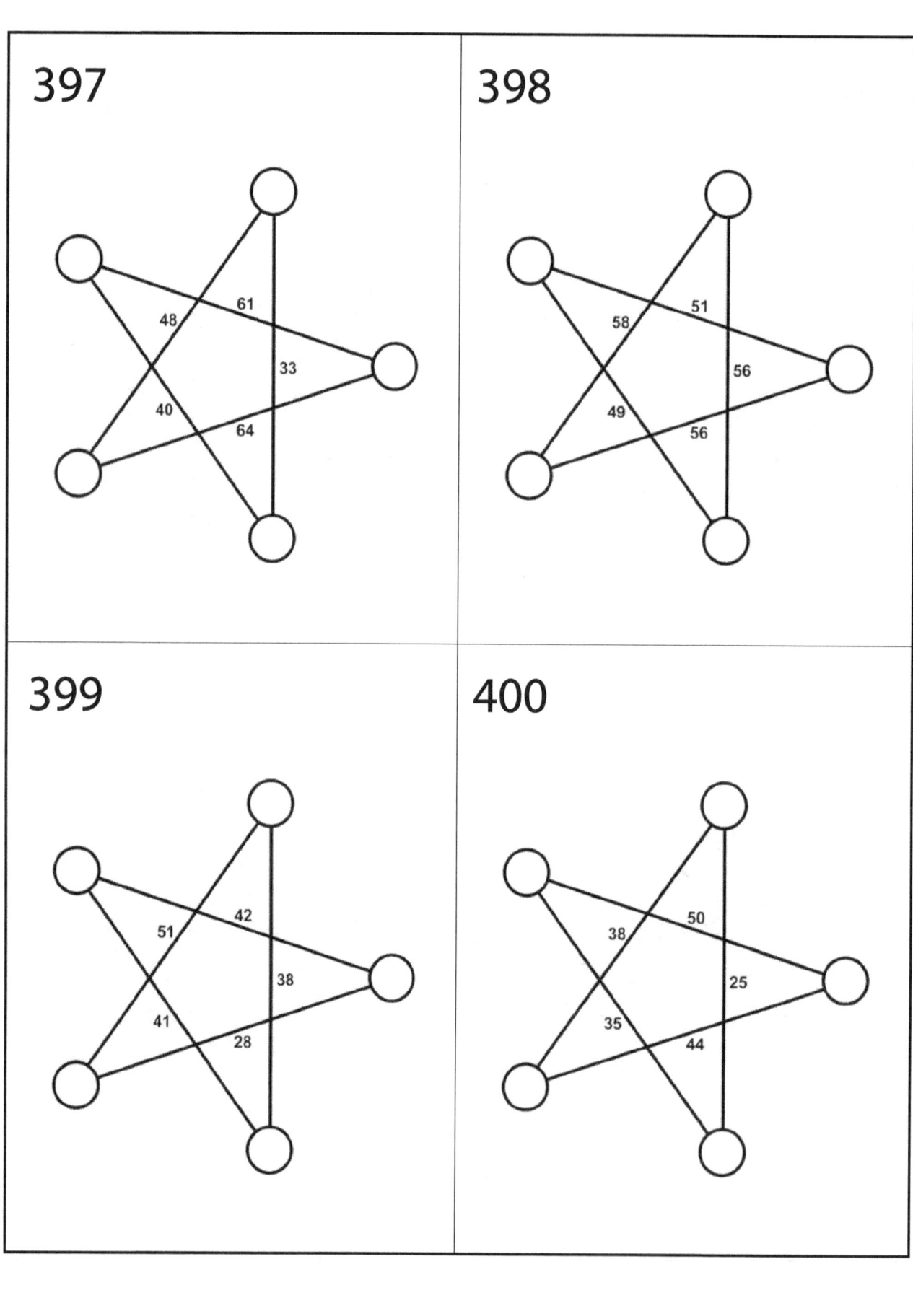

www.ingramcontent.com/pod-product-compliance
Lightning Source LLC
Chambersburg PA
CBHW081119240526
45470CB00019B/2629